NANOSCALE FLOW

Advances, Modeling, and Applications

NANOSCALE FLOW

Advances, Modeling, and Applications

EDITED BY
Sarhan M. Musa

CRC Press
Taylor & Francis Group
Boca Raton London New York

CRC Press is an imprint of the
Taylor & Francis Group, an **informa** business

MATLAB® is a trademark of The MathWorks, Inc. and is used with permission. The MathWorks does not warrant the accuracy of the text or exercises in this book. This book's use or discussion of MATLAB® software or related products does not constitute endorsement or sponsorship by The MathWorks of a particular pedagogical approach or particular use of the MATLAB® software.

CRC Press
Taylor & Francis Group
6000 Broken Sound Parkway NW, Suite 300
Boca Raton, FL 33487-2742

First issued in paperback 2017

© 2015 by Taylor & Francis Group, LLC
CRC Press is an imprint of Taylor & Francis Group, an Informa business

No claim to original U.S. Government works

ISBN-13: 978-1-4822-3380-3 (hbk)
ISBN-13: 978-1-138-74921-4 (pbk)

**Visit the Taylor & Francis Web site at
http://www.taylorandfrancis.com**

**and the CRC Press Web site at
http://www.crcpress.com**

Dedicated to my late father, Mahmoud, my mother, Fatmeh, and my wife, Lama.

Contents

Preface..ix
Acknowledgments...xi
Editor...xiii
Contributors...xv

1. Boiling Heat Transfer and Critical Heat Flux
   Phenomena of Nanofluids 1
   Lixin Cheng

2. Modeling for Heat Transfer of Nanofluids
   Using a Fractal Approach.. 31
   Boqi Xiao

3. Thermal Conductivity Enhancement in
   Nanofluids Measured with a Hot-Wire Calorimeter............................79
   Catalina Vélez, José M. Ortiz de Zárate, and Mohamed Khayet

4. Two-Phase Laminar Mixed Convection
   Al₂O₃–Water Nanofluid in Elliptic Duct... 101
   Buddakkagari Vasu and Rama Subba Reddy Gorla

5. Nanooncology: Molecular Imaging, Omics, and Nanoscale
   Flow-Mediated Medicine Tumors Strategies.. 121
   Tannaz Farrahi, Tri Quang, Keerthi Srivastav Valluru,
   Suman Shrestha, George Livanos, Yinan Li, Aditi Deshpande,
   Michalis Zervakis, and George C. Giakos

6. Nanoscale Flow Application in Medicine 171
   Viroj Wiwanitkit

Appendix A: Material and Physical Constants 205

Appendix B: Photon Equations, Index of Refraction,
            Electromagnetic Spectrum, and Wavelengths
            of Commercial Lasers 209

Appendix C: Symbols, Formulas, and Periodic Table................................215

Index...235

Preface

Understanding the behavior of flow at the nanoscale has been of great interest in recent years. As a typical flow in the reduced-size fluid mechanics system, the nanochannel flow embodies a series of special properties and has attracted much research attention. The understanding of the physical properties and dynamical behavior of nanochannel flows is important for the theoretical study of fluid dynamics and many engineering applications in physics, chemistry, medicine, and electronics. The flows inside nanoscale pores are also important due to their highly beneficial drag and heat transfer properties. This book provides very strong recent discovery and research in the area of the multidisciplinary principles of nanoscale flow advances, modeling, and applications by inventors and researchers from around the world.

This book contains six chapters and three appendices. Chapter 1 provides a comprehensive review of the current status of research on nucleate pool boiling heat transfer, flow boiling heat transfer, and critical heat flux (CHF) phenomena of nanofluids.

Chapter 2 presents two novel fractal models for pool boiling heat transfer of nanofluids, including subcooled pool boiling and nucleate pool boiling.

Chapter 3 provides thermal conductivity enhancement in nanofluids measured with a hot-wire calorimeter.

Chapter 4 presents two-phase laminar mixed convection AL_2O_3–water nanofluid in elliptic duct.

Chapter 5 presents the principles of molecular and omics imaging and spectroscopy techniques for cancer detection. The authors analyze fluid dynamics modeling of the tumor vasculature and drug transport. Also, they study the properties of nanoscale particles and their impact on the diagnosis, therapeutics, and theranostics.

Chapter 6 presents a brief background and review on medical nanoscale flow applications.

The book concludes with the appendices. Appendix A presents common material and physical constants, with the consideration that the constants' values varied from one published source to another because there are many varieties of materials; conductivity is sensitive to temperature, impurities, and moisture content; and the dependence of relative permittivity and permeability on temperature and humidity.

Appendix B provides equations for photon energy, frequency, wavelength, and electromagnetic spectrum, including the approximation of common optical wavelength ranges of light. In addition, it contains a figure illustrating the wavelengths of commercially available lasers.

Appendix C provides common symbols, useful mathematical formulas, and the periodic table.

Sarhan M. Musa
Houston, Texas

MATLAB® is a registered trademark of The MathWorks, Inc. For product information, please contact:

The MathWorks, Inc.
3 Apple Hill Drive
Natick, MA 01760-2098 USA
Tel: 508-647-7000
Fax: 508-647-7001
E-mail: info@mathworks.com
Web: www.mathworks.com

Acknowledgments

My sincere appreciation and gratitude to all the book's contributors. Thanks to Brian Gaskin and James Gaskin for their wonderful hearts and for being great American neighbors. It is my pleasure to acknowledge the outstanding help and support of the team at Taylor & Francis Group/CRC Press in preparing this book, especially Nora Konopka, Michele Smith, Kari Budyk, and Todd Perry.

I thank Professors John Burghduff and Mary Jane Ferguson for their support and understanding and for being great friends. Thanks also to Dr. Korsh Jafarnia for taking good care of my mother's health during the course of this project.

I thank Dr. Kendall T. Harris, my college dean, for his constant support. Finally, the book would never have seen the light of day if not for the constant support, love, and patience of my family.

Editor

Sarhan M. Musa, PhD, is associate professor in the Department of Engineering Technology, Roy G. Perry College of Engineering, at Prairie View A&M University, Texas. He has been director of the Prairie View Networking Academy, Texas, since 2004. Dr. Musa has published more than a hundred papers in peer-reviewed journals and conferences. He is a frequent invited speaker in computational nanotechnology, has consulted for multiple organizations nationally and internationally, and has written and edited several books, including *Computational Nanotechnology Modeling and Applications with MATLAB®*. Dr. Musa is a senior member of the Institute of Electrical and Electronics Engineers (IEEE) and is also an LTD Sprint and Boeing Welliver fellow.

Contributors

Lixin Cheng
Department of Engineering
Aarhus University
Aarhus, Denmark

Aditi Deshpande
Department of Biomedical
    Engineering
The University of Akron
Akron, Ohio

Tannaz Farrahi
Charles L. Brown Department of
    Electrical Engineering
University of Virginia
Charlottesville, Virginia

George C. Giakos
Department of Electrical and
    Computer Engineering
Manhattan College
Riverdale, New York

Rama Subba Reddy Gorla
Department of Mechanical
    Engineering
Cleveland State University
Cleveland, Ohio

Mohamed Khayet
Department of Applied Physics I
University Complutense of Madrid
Madrid, Spain

Yinan Li
Department of Electrical and
    Computer Engineering
The University of Akron
Akron, Ohio

George Livanos
Department of Electronic
    Engineering and Computer
    Science
Technical University of Crete
Chania, Greece

José M. Ortiz de Zárate
Department of Applied Physics I
University Complutense of Madrid
Madrid, Spain

Tri Quang
Department of Biomedical
    Engineering
The University of Akron
Akron, Ohio

Suman Shrestha
Department of Radiology
University of Massachusetts
    Medical School
Worcester, Massachusetts

Keerthi Srivastav Valluru
Department of Electrical and
    Computer Engineering
The University of Akron
Akron, Ohio

Buddakkagari Vasu
Department of Mathematics
Motilal Nehru National Institute of
    Technology Allahabad
Allahabad, Uttar Pradesh, India

Catalina Vélez
Department of Applied Physics I
University Complutense of Madrid
Madrid, Spain

Viroj Wiwanitkit
Department of Tropical Medicine
    (retired)
Hainan Medical University
Haikou, Hainan, People's Republic
    of China

Boqi Xiao
School of Mechanical and Electrical
    Engineering
Sanming University
Sanming, Fujian, People's
    Republic of China

Michalis Zervakis
Department of Electronic
    Engineering and Computer
    Science
Technical University of Crete
Chania, Greece

1

Boiling Heat Transfer and Critical Heat Flux Phenomena of Nanofluids

Lixin Cheng

CONTENTS

1.1 Introduction 1
1.2 Physical Properties of Nanofluids 3
    1.2.1 Thermal Conductivity of Nanofluids 3
    1.2.2 Viscosity of Nanofluids 6
    1.2.3 Other Thermophysical Properties of Nanofluids 8
1.3 Experimental Studies on Boiling Heat Transfer and Critical Heat
    Flux of Nanofluids 9
    1.3.1 Experimental Studies on Nucleate Pool Boiling and
        CHF of Nanofluids 9
        1.3.1.1 Experimental Studies on Nucleate Pool Boiling
               of Nanofluids 10
        1.3.1.2 Experimental Studies on CHF of Nucleate
               Pool Boiling of Nanofluids 13
    1.3.2 Studies on Flow Boiling Heat Transfer and
        CHF of Nanofluids 14
1.4 Boiling Heat Transfer and CHF Mechanisms of Nanofluids 16
1.5 Challenges and Future Research Needs 21
1.6 Concluding Remarks 22
References 23

1.1 Introduction

Heat transfer nanofluids were first reported by Choi [1] of the Argonne National Laboratory, United States, in 1995. It has been demonstrated that nanofluids can have significantly better heat transfer characteristics than the base fluids. The following key features of nanofluids have been found [2–6]: (1) they have larger thermal conductivities compared to conventional fluids, (2) they have a strongly nonlinear temperature dependency on the effective thermal conductivity, (3) they enhance or diminish heat transfer

1

in single-phase flow, (4) they enhance or reduce nucleate pool boiling heat transfer, and (5) they yield higher critical heat fluxes (CHFs) under pool boiling conditions. Furthermore, some contradictory trends in heat transfer have been reported in the literature [2–5]. In general, research on heat transfer performance, heat transfer enhancement mechanisms, and nanofluid applications is still in its primary stage. The use of nanofluids appears promising in several aspects of thermal physics but still faces several challenges: (1) the lack of agreement between experimental results from different research groups and (2) the lack of theoretical understanding of the underlying mechanisms with respect to nanoparticles. Especially, as new research frontiers of nanotechnology, there are great challenges of research and applications of two-phase flow and thermal physics of nanofluids [3,4].

Both nucleate pool boiling and flow boiling processes are frequently encountered in many industrial applications such as distillation columns, boiler tubes, evaporators, chemical reactors, refrigeration, air-conditioning and heat-pumping systems, nuclear reactors, cooling of electronic components, and so on. Quite a number of studies on the nucleate pool boiling and flow boiling heat transfer phenomena, including bubble behaviors, flow regimes, heat transfer, CHF, and two-phase pressure drop and the relevant prediction methods, have been performed in large spaces and macroscale channels over the past decades [7]. In recent years, flow boiling in microchannels has become one of the *hottest* research topics in heat transfer as a highly efficient cooling technology, as it has numerous advantages of high heat transfer performance, chip temperature uniformity, hot spots cooling capability, and more [8–12]. Enhancement of boiling and CHF is important in improving energy efficiency and operation safety in various applications [13,14]. One such method is to use nanofluids as heat transfer fluids in various applications to enhance boiling heat transfer and CHF. In recent years, a new research frontier of nanofluids two-phase flow and heat transfer is under rapid development. It seems that nanofluids may significantly enhance CHF in both nucleate pool boiling and flow boiling processes as they might be important methods to enhance CHF and have potential applications in various industries. It is thus recommended that further research be necessary to investigate the CHF phenomena of nanofluids in microscale channels and confined spaces in the future, as such researches are rare so far [4].

Nanofluids are engineered colloids made of a base fluid and nanoparticles (1–100 nm). Common base fluids include water, organic liquids (e.g., ethylene, tri-ethylene-glycols, and refrigerants), oils and lubricants, biofluids, polymeric solutions, and other common liquids. Materials commonly used as nanoparticles include chemically stable metals (e.g., gold and copper), metal oxides (e.g., alumina, silica, zirconia, and titania), oxide ceramics (e.g., Al_2O_3 and CuO), metal carbides (e.g., SiC), metal nitrides (e.g., AlN and SiN), carbon in various forms (e.g., diamond, graphite, carbon nanotubes (CNTs), and fullerene), and functionalized nanoparticles. The use of nanofluids is a new research frontier related to nanotechnology and has found a wide range of potential applications. According to the application, nanofluids are classified as heat transfer nanofluids, tribological nanofluids, surfactant and coating nanofluids, chemical nanofluids, process/extraction nanofluids, environmental (pollution cleaning) nanofluids, bio- and pharmaceutical nanofluids and medical nanofluids (drug delivery, functional, and tissue–cell interaction). As a fluid class, nanofluids have a unique feature that is quite different from those of conventional solid–liquid mixtures in which millimeter- and/or micrometer-sized particles are added. Such particles settle rapidly, clog flow channels, erode pipelines, and cause severe pressure drops. All these shortcomings prohibit the application of conventional solid–liquid mixtures in microchannels while nanofluids instead can be used in microscale heat transfer. Furthermore, compared to nucleate pool boiling enhancement by the addition of surfactants, nanofluids can enhance the CHF while surfactants normally do not [2–6,14]. Thus, nanofluids appear promising as coolants for dissipating very high heat fluxes in various applications.

Researchers have given much more attention rather to the thermal conductivity of nanofluids than their heat transfer characteristics. Most of the available heat transfer studies are related to single-phase flows, and some are related to nucleate pool boiling. However, the study of flow boiling and the two-phase flow of nanofluids is very limited in the literature so far. None of the available reviews [2–6] has specifically mentioned this new important research frontier, although all have presented studies of nucleate pool boiling heat transfer of nanofluids with very brief descriptions. Therefore, the present review seeks to critically confront and summarize the state of the art and understanding of boiling heat transfer and CHF characteristics of nanofluids and to identify particular areas requiring further study.

1.2 Physical Properties of Nanofluids

Boiling heat transfer and CHF characteristics of nanofluids depend on accurate thermophysical properties such as specific heat, latent heat, density, surface tension, and so on. However, so far, most studies on nanofluid thermal properties have focused on thermal conductivity and limited studies on viscosity. Other physical properties and their methods are rare in the literature, but their studies must be performed in the future.

1.2.1 Thermal Conductivity of Nanofluids

Solids have thermal conductivities that are orders of magnitude larger than those of conventional heat transfer fluids as shown in Table 1.1. By suspending nanoparticles in conventional heat transfer fluids, the heat transfer performance of the fluids can be significantly improved [2–6].

TABLE 1.1

Thermal Conductivities of Various Solids and Liquids at
Room Temperature

Material	Form	Thermal Conductivity (W/mK)
Carbon	Nanotubes	1800–6600
	Diamond	2300
	Graphite	110–190
	Fullerenes film	0.4
Metallic solids (pure)	Silver	429
	Copper	401
	Nickel	237
Nonmetallic solids	Silicon	148
Metallic liquids	Aluminum	40
	Sodium at 644 K	72.3
Others	Water	0.613
	Ethylene glycol	0.253
	Engine oil	0.145
	R134a	0.0811

The main reasons may be as follows: (1) The suspended nanoparticles increase the surface area and the heat capacity of the fluid, (2) the suspended nanoparticles increase the effective (or apparent) thermal conductivity of the fluid, (3) the interactions and collisions among particles, fluid, and the flow passage surface are intensified, (4) the mixing and turbulence of the fluid are intensified, and (5) the dispersion of nanoparticles enable a more uniform temperature distribution in the fluid.

Thermal conductivity enhancement for a variety of nanoparticles/base fluids has also been reported [15–23], of which a few are mentioned here. Eastman et al. [15] reported that a small amount (about 0.3% by volume fraction) of copper nanoparticles of mean diameter <10 nm in ethylene glycol increased the fluid's inherently poor thermal conductivity by 40%. Substantially increased thermal conductivities of nanofluids containing a small amount of metal such as Cu and Fe or metal oxide such as SiO_2, Al_2O_3, WO_3, TiO_2, and CuO have been reported in the literature [4]. Experimental results have shown that these nanofluids have substantially higher thermal conductivities than the same liquids without nanoparticles. The nanoparticle thermal conductivity increases with the nanoparticle volume fraction. In general, metallic nanofluids show much more dramatic enhancements than metallic oxide nanofluids. Furthermore, nanofluid thermal conductivities are also strongly dependent on temperature [18,23]. Particle size, shape, and volume concentration also influence the thermal conductivity of nanofluids [16–18,23]. Hong et al. [20] concluded that the thermal conductivity of an Fe nanofluid is increased nonlinearly up to 18% as the volume fraction of particle is increased to 0.55 vol.%. Comparing Fe nanofluids with Cu nano-fluids, they found that the suspension of highly thermally conductive mate-rials is not always effective.

In recent years, CNTs have attracted much attention because of their unique structure and remarkable mechanical and electrical properties [24,25]. Recent studies reveal that CNTs have unusually high thermal conductivity of up to 6000 W/mK compared to 0.08 W/mK of a liquid refrigerant [25–28]. CNTs can thus apparently further enhance the ther-mal conductivity of nanofluids [29–34]. In addition, CNTs have a very high aspect ratio [3,4]. CNTs from a highly entangled fiber network are not very mobile, as demonstrated by viscosity measurements, and thus, their effect on the thermal transport in fluid suspensions is expected to be similar to that of polymer composites. The first reported work on a single-walled carbon nanotube (SWCNT)-polymer epoxy composite by Biercuk et al. [25] demonstrated a 70% increase in thermal conductivity at 40 K, rising to 125% at room temperature with 1 wt.% nanotube loading. They also observed that thermal conductivity increased with increasing tempera-ture. Hone et al. [26] reported that the thermal conductivity of SWCNTs was linear in temperature from 7 to 25 K, increased in slope between 24 and 40 K, and then rose monotonically with temperature to above room temperature. Berber et al. [27] reported an unusually high thermal conduc-tivity of CNTs, reaching 6600 W/mK at room temperature. Kim et al. [28] reported that the thermal conductivity of individual multiwalled nano-tubes reached 3000 W/mK at room temperature. Choi et al. [29] measured thermal conductivities of oil suspensions containing multiwalled carbon nanotubes (MWCNTs) up to 1 vol.% loading and found similar behavior, in this case, a 160% enhancement.

Dispersion of a small amount of nanotubes produces a remarkable change in the thermal conductivity of the base fluid (up to 259% at 1 vol.%) [29]. The thermal conductivity of nanotube suspensions (solid circles) is of the order of magnitude greater than predicted by the existing models (dotted lines). The measured thermal conductivity of nanotube suspensions is nonlinear with nanotube volume fraction, while theoretical predictions show a lin-ear relationship (inset), which is thus a significant contradiction to what is expected. With the increasing concentration, the thermal conductivity is greatly increased. Xie et al. [30] found a 10%–20% enhancement of effective thermal conductivities of CNT suspensions in distilled water and ethylene glycol. Several other studies also reported the enhancement of thermal con-ductivity of CNT nanofluids [33,34].

From a theoretical aspect, because of the absence of a theory for thermal conductivities of nanofluids, the existing models developed for conven-tional solid/liquid systems have been used to estimate the effective con-ductivities of nanofluids [2–6]. For example, the Hamilton and Crosser [35] model has been applied to nanofluids. However, measured thermal con-ductivities are substantially greater than theoretical predictions [21,29–31].

Furthermore, a number of investigations have been conducted to identify the possible mechanisms that contribute to the enhanced effective thermal conductivity of nanoparticle suspensions. The Brownian motion of the nanoparticles in these suspensions is one of the potential contributors to this enhancement [19,21]. A number of theoretical studies have accounted for the higher thermal conductivity considering other factors [19,21,23,36,37]. For example, Jang and Choi [19] devised a theoretical model that accounts for the fundamental role of the dynamics of nanoparticles in nanofluids. The model not only captures the concentration and temperature-dependent effects but also predicts a strong size-dependent influence. With respect to heat transfer enhancement, Jang and Choi [21] proposed four potential mechanisms for the anomalous increase in nanofluid heat transfer: the Brownian motion of nanoparticles, ballistic phonon transport inside nanoparticles, interfacial layering of liquid molecules, and nanoparticle clustering. However, the development of theoretical thermal conductivity model based on such mechanisms is still a challenge. At present, the fundamental mechanisms are not yet well understood, and no concrete conclusions have been reached that prove which is/are the controlling mechanisms.

Further research efforts are needed to develop a suitable model to predict the thermal conductivity of nanofluids and should take into account the important molecular and nanomechanisms that are responsible for enhancing the thermal conductivity of nanofluids. In fact, fundamental studies should be performed to provide improved insight into the mechanisms of the thermal conductivity of nanofluids, as have been pointed out by Cheng and Liu [3] and Cheng [6]. Furthermore, since nanoparticles can form nano- or microstructures, it would seem that the thermal conductivity of such a nanofluid under static conditions could be quite different under flow conditions.

1.2.2 Viscosity of Nanofluids

A few studies have addressed the viscous properties of nanofluids as summarized by Cheng and Liu [3] and Cheng [6]. The viscosity of nanofluids is normally much higher than that of their base fluids. The viscosity is a strong function of temperature and the volumetric concentration. Furthermore, a particle-size effect seems to be important only for sufficiently high particle fractions.

Kulkarni et al. [38] conducted an experimental investigation on the rheological behavior of copper oxide nanoparticles dispersed in a 60:40 propylene glycol and water mixture, with particle volumetric concentrations from 0% to 6% at temperatures from −35°C to 50°C. Their results showed that these nanofluids exhibited a Newtonian fluid behavior. Kulkarni et al. [39] studied the rheological property of CuO–water nanofluids with volumetric concentrations of 5%–15% at temperatures from 278 to 323 K. Their experimental results showed that these fluids behave as time-independent, shear thinning, pseudoplastic fluids. They also proposed a new correlation to predict the viscosity of these nanofluids as a function of temperature and volumetric concentration based on their own data.

Nguyen et al. [40] studied the effect of temperature and particle volume concentration on the dynamic viscosity for water–Al_2O_3 nanofluids at temperatures from 22°C to 75°C. They reported a hysteresis phenomenon on viscosity. Their experimental data showed that for a given particle volume concentration, there is a critical temperature beyond which nanofluid viscous behavior becomes drastically altered. If a fluid sample is heated beyond such a critical temperature, a striking increase of viscosity occurs. If it is cooled after being heated beyond this critical temperature, then a hysteresis phenomenon can occur as shown in Figure 1.1. Such an intriguing hysteresis phenomenon still remains poorly understood. Furthermore, the critical temperature was found to be strongly dependent on both particle fraction and size.

Ding et al. [34] have reported that the viscosity of CNT nanofluids increased with increasing concentration and decreasing temperature. Figure 1.2 shows their viscosity measurements for CNT nanofluids at pH = 6. A shear thinning behavior was also observed. Their nanofluids showed a nonlinear effect at high shear rates, which is actually non-Newtonian fluid behavior.

So far, no systematic theory or generalized model is available to predict the viscosity of nanofluids. Further experimental study is needed to expand the database while fundamental investigations on fluid/particle surface interactions should be made as a prerequisite for the development of theoretical models. Both Newtonian and non-Newtonian models should be developed.

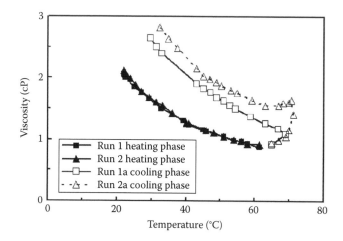

FIGURE 1.1

Hysteresis observed for water–Al_2O_3–47 nm, 7% particle volume fraction. (Reprinted from *Int. J. Therm. Sci.*, 47, Nguyen, C.T., Desgranges, F., Galanis, N., Roy, G., Maré, T., Boucher, S., and Minsta, H.A., Viscosity data for Al_2O_3–water nanofluid—Hysteresis: Is heat transfer enhancement using nanofluids reliable?, 103–111, Copyright 2008, with permission from Elsevier.)

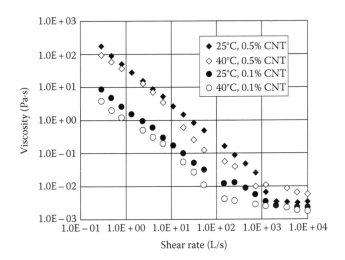

FIGURE 1.2
Viscosity of CNT nanofluids (pH = 6). (Reprinted from *Int. J. Heat Mass Transfer*, 49, Ding, Y., Alias, H., Wen, D., and Williams, R.A., Heat transfer of aqueous suspensions of carbon nanotubes, 240–250, Copyright 2005, with permission from Elsevier.)

1.2.3 Other Thermophysical Properties of Nanofluids

Boiling heat transfer and CHF characteristics also depend on other physical properties such as specific heat, latent heat, density, surface tension, and so on. Apparently, no such studies are available in the open literature on these properties but should be performed in the future. Some researchers have dealt with specific heat and density of nanofluids using various mixing rules. However, further measurements should be done to verify their applicability to nanofluids.

Surface tension is a very important parameter in boiling and two-phase flow heat transfer phenomena. However, so far, there has only been one study by Xue et al. [41] presenting surface tension data for a CNT nanofluid. With the addition of CNTs to water, the surface tension increases by about 14% compared to pure water at the same temperature. This is contrary to the effect arising from surface tension reduction by the addition of a surfactant in water as shown in Figure 1.3 by Cheng [14]. The higher surface tension of CNT suspensions could modify the mechanisms controlling nucleate boiling, flow boiling, CHF, and flow pattern transitions, which might be characterized by the formation of larger-sized bubbles with diminished departure frequencies and an increased tendency to coalesce. Furthermore, two-phase pressure drops may be affected as well. However, there is only this one study available so far. More measurements are needed to build a database for surface tension and other physical properties for nanofluids.

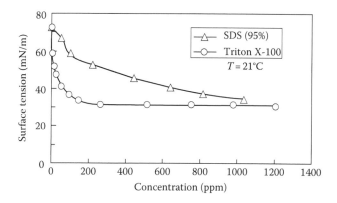

FIGURE 1.3
Variation of the measured equilibrium surface tension versus surfactant concentration. (Reprinted from *Int. J. Heat Mass Transfer*, 50, Cheng, L., Mewes, D., and Luke, A., Boiling phenomena with surfactants and polymeric additives: A state-of-the-art review, 2744–2771, Copyright 2007, with permission from Elsevier.)

Recently, several studies have shown that the specific heat can be enhanced using nanoparticles [42–44], which is of significance for boiling heat transfer and CHF phenomena. Such studies are very rare and need to be further investigated experimentally and theoretically.

1.3 Experimental Studies on Boiling Heat Transfer and Critical Heat Flux of Nanofluids

The study of nanofluid boiling heat transfer and CHF phenomena is very limited as pointed out by Cheng et al. [4] and Cheng and Liu [3]. However, the number of publications on this subject has greatly increased in recent years, which indicates that the importance of this field has now been recognized but still needs to be further investigated as the number of publications is still inadequate. In the following, experimental studies on several related topics such as nucleate pool boiling, CHF, and flow boiling heat transfer are reviewed.

1.3.1 Experimental Studies on Nucleate Pool Boiling and CHF of Nanofluids

A number of experimental studies on nucleate pool boiling heat transfer and CHF have been conducted. The experimental data on nucleate pool boiling are conflicting with some studies, showing a decrease or no change in nucleate boiling heat transfer with the addition of nanoparticles, while some show an increase. However, all the available studies are in agreement in showing an enhancement in CHF at pool boiling conditions. In this section, an overall review on nucleate pool boiling heat transfer and CHF is presented. The following results must be put in proper perspective. For instance, nucleate pool boiling data are often measured with about ±10% errors and experimental data from independent studies on the same pure fluid often disagree by 30%–50% or more.

1.3.1.1 Experimental Studies on Nucleate Pool Boiling of Nanofluids

Witharana [45] measured heat transfer coefficients of Au (unspecified size)–water, SiO_2 (30 nm)–water, and SiO_2–ethylene glycol nanofluids under pool boiling conditions. Results for Au–water nanofluids showed the nanofluid heat transfer coefficients were higher than those of pure water and also increased with increasing gold particle concentration. The enhancement of heat transfer was only about 11% at the intermediate heat fluxes (3 W/cm²) and 21% at a higher heat flux (4 W/cm²). However, the SiO_2–water and SiO_2–ethylene glycol nanofluids depicted a decrease in their heat transfer coefficients. These contradictory behaviors were not explained in the study.

Yang and Liu [46] conducted an experimental investigation on nucleate pool boiling heat transfer performance of refrigerant R-141b with and without nanosized Au particles on a horizontal plain tube. Three concentrations of 0.09, 0.45, and 1 vol.% were used in their experiments. For R-141b with 0.09 vol.% nanoparticles, there was no significant effect on pool boiling heat transfer performance for such a low concentration of nanoparticles. Their test results show that the heat transfer coefficients for pure R-141b agree very well with those predicted by Cooper [47] correlation while the boiling heat transfer coefficients increase with increasing nanoparticle concentration. At a particle concentration of 1.0 vol.%, the heat transfer coefficient is more than twice higher than those without nanoparticles. The addition of nanoparticles significantly increased R-141b boiling heat transfer behavior. Their results agree with those by Wen and Ding [48] but are in contradiction to those of Das et al. [49] and Bang and Chang [50], who observed decrease in boiling heat transfer coefficients due to the presence of nanoparticles. Furthermore, they repeated their measurements of heat transfer coefficients four times with 5-day intervals as shown in Figure 1.2. It shows that the measured boiling heat transfer coefficients decreased for each test and finally reached close to those without nanoparticles. They attribute this to the trapped particles on surface and reduced the number of activation nucleation sites. The SPM investigation shows that the test tube surface roughness decreased from 0.317 μm before the boiling test to 0.162 μm after the test. Further investigation by transmission electron microscopy (TEM) and dynamic light scattering particle analyzer shows that the nanoparticles aggregated from the size of 3 nm before the test to 110 nm after the test. From their careful study, it can be concluded that the nanosized Au particles are able to significantly increase pool boiling heat transfer of refrigerant R-141b on plain tube surface. However, tube surface roughness and particle size changed after boiling test. Both of these effects degrade the boiling heat transfer coefficients.

Several other studies have shown that nanoparticles do not enhance nucleate pool boiling heat transfer or decrease it. Li et al. [51] conducted experiments to investigate pool boiling heat transfer characteristics of a CuO/water nanofluid and found that heat transfer deteriorated. They attributed this to the decrease in active nucleation sites caused by nanoparticle sedimentation on the boiling surface, based on observations. Das et al. [49,52] carried out an experimental study on pool boiling characteristics of Al$_2$O$_3$–water nanofluids under atmospheric conditions. Figure 1.4 shows their experimental results on a smooth heater and a rough heater [49]. They found that the nanoparticles degraded the boiling performance. They speculated that the deterioration in boiling performance was not due to a change in fluid property but due to the change in surface wetting characteristics because of the entrapment of nanoparticles in the surface cavities. You et al. [53] also reported deterioration in nucleate pool boiling heat

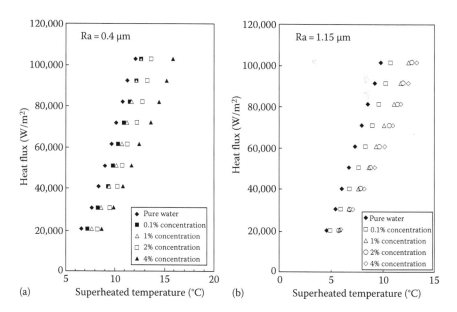

FIGURE 1.4
(a) Pool boiling characteristics of Al$_2$O$_2$–water nanofluids at different particle concentrations on a smooth heater. (b) Pool boiling characteristics of Al$_2$O$_2$–water nanofluids at different particle concentrations on a rough heater. (Reprinted from *Int. J. Heat Mass Transfer*, 46, Das, S.K., Putra, N., and Roetzel, W., Pool boiling characteristics of nanofluids, 851–862, Copyright 2003, with permission from Elsevier.)

transfer for Al_2O_3–water nanofluids. Kim et al. [54] found that heat transfer coefficients of Al_2O_3–water nanofluids remained unchanged compared to those of water. Bang and Chang [50] studied boiling heat transfer using Al_2O_3–water nanofluids on a horizontal smooth surface. They showed that nanofluids have poorer heat transfer coefficients compared to pure water in natural convection as well as in nucleate boiling. Vassallo et al. [55] compared the heat transfer performance of pure water, silica nanosolutions, and silica microsolutions under atmospheric pressure boiling on a 0.4 mm NiCr wire submerged in each solution. Their results also showed no appreciable differences in nucleate boiling heat transfer. A thick (0.15–0.2 mm) silica coating was observed to form on their wire heater. They speculated that the roughness of the solid substrate might be responsible for the observed results. The additional thermal resistance of the silica could also have played a role. Prakash Narayan et al. [56] studied the effect of surface orientation on nucleate pool boiling of Al_2O_3–water nanofluids and found that heat transfer deteriorated for all test cases. A significant effect of surface orientation on heat transfer was found, where a horizontal orientation gave the best heat transfer and the heater surface at an inclination of 45° gave the worst heat transfer.

On the contrary, several studies have shown that nanoparticles can enhance nucleate pool boiling heat transfer. Tu et al. [57] found significant enhancement in pool boiling heat transfer for an Al_2O_3–water nanofluid, up to 64% for a small fraction of nanoparticles. Wen and Ding [48] conducted experiments on nucleate pool boiling using γ-Al_2O_3–water nanofluids. They found that the presence of alumina in the nanofluid enhanced heat transfer significantly, by up to 40% for a 1.25 wt.% concentration of the nanoparticles. Wen et al. [58] conducted experiments on pool boiling of TiO_2–water nanofluids and showed that heat transfer increased by up to 50% at a concentration of 0.7 vol.%. Chopkar et al. [59] reported that ZrO_2–water can enhance nucleate boiling heat transfer at low particle volumetric concentrations but heat transfer decreases with a further increase in concentration. They also mentioned that the addition of a surfactant to the nanofluids drastically decrease heat transfer whereas surfactants often increase nucleate boiling heat transfer.

More recently, CNTs have been used to enhance nucleate boiling heat transfer. Park and Jung [60] studied the effect of CNTs on nucleate boiling heat transfer of two halocarbon refrigerants (R123 and R134a). One volume percent of CNTs was added to the refrigerants, and they found that CNTs increased nucleate boiling heat transfer coefficients for both refrigerants. Figure 1.5 shows their heat transfer coefficients of R134a with and without CNTs [60]. In particular, enhancements up to 36.6% were observed at low heat fluxes. With increasing heat flux, however, the enhancement diminished due to more vigorous bubble generation according to their visual observations. In addition, no deposit of the particles on their heat transfer surface was observed in their study.

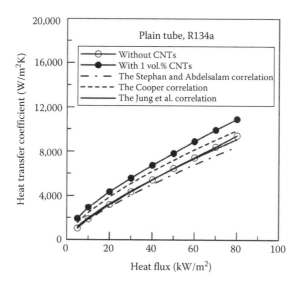

FIGURE 1.5

Boiling heat transfer coefficients with 1 vol.% CNTs for R134a. (Reprinted from *Energy Build.*, 39, Park, K.J., and Jung, D., Boiling heat transfer enhancement with carbon nanotubes for refrigerants used in building air-conditioning, 1061–1064, Copyright 2007, with permission from Elsevier.)

1.3.1.2 Experimental Studies on CHF of Nucleate Pool Boiling of Nanofluids

All the available experimental studies mentioned on CHF have shown some enhancement of CHF. The maximum increase in CHF with a nanofluid can be three times more compared to that of the base fluid [55].

You et al. [53] found drastic enhancement in CHF for Al_2O_3–water nanofluids. Their experimental results for CHF in pool boiling shows that up to threefold enhancement can be achieved. They also performed a visualization study and found that the average size of departing bubbles increased but the frequency of departing bubble decreased. They concluded that the unusual CHF enhancement of nanofluids could not be explained by any existing CHF model; that is, no pool boiling CHF model includes thermal conductivity or liquid viscosity and hence cannot explain this phenomenon, and the enhancement on liquid-to-vapor phase change was not related to the increased thermal conductivity. They also reported enhancement in CHF for both horizontal and vertical surface orientations with the nanofluids. Noting a change of the roughness of the heater surface before and after their experiments, they hypothesized that the reason for the increase in CHF might be due to a surface coating formed on the heater with nanoparticles.

Kim et al. [54] studied the effect of nanoparticles on CHF enhancement in pool boiling of Al_2O_3–water and found that CHF was improved by up to 200% at a concentration of 0.01 g/L. They also observed that the size of departing bubbles increased and the bubble frequency decreased significantly in nanofluids compared to those in pure water. The orientation of the heater surface had a great effect on CHF. Kim et al. [61,62] also reported that a significant enhancement in CHF was achieved for Al_2O_3–water, ZrO_2–water, and SiO_2–water nanofluids at concentrations less than 0.1 vol.%. They noted that a porous layer of nanoparticles was formed on the heat transfer surface and this layer apparently improved the surface wettability significantly, which may explain a plausible mechanism of CHF enhancement. Kim et al. [63,64], and Kim and Kim [65] found that CHF enhancement was achieved for Al_2O_3–water and TiO_2–water nanofluids at concentrations from 0.005 to 0.1 vol.% with a maximum increase of about 100%. They also found that nanoparticles coated their heat transfer surfaces and thus apparently enhanced the surface wettability. Hence, if augmentation is the result of the coating, one could conclude that it is more effective to directly coat the surface rather than use nanoparticles in the fluid to do it.

Tu et al. [57] again found that CHF enhancement was achieved for Al_2O_3–water nanofluids. Vassallo et al. [55] reported a marked increase in CHF for SiO_2–water nanoparticles with a maximum value of up to three times compared to that of pure water. Milanova and Kumar [66] found that the CHF of ionic solutions with SiO_2 nanoparticles was enhanced up to three times compared to that of conventional fluids. Nanofluids in a strong electrolyte, that is, in a high ionic concentration, yielded higher CHF than buffer solutions. Xue et al. [67] conducted experiments on CNTs–water nanofluids and found that CHF, transition boiling, and the minimum heat flux in film boiling were enhanced.

1.3.2 Studies on Flow Boiling Heat Transfer and CHF of Nanofluids

So far, the limited studies on flow boiling heat transfer of nanofluids have shown contradictory experimental results. The available studies are the thermal performance of thermosyphons, heat transfer in heat pipes, spray cooling of nanofluids, and flow boiling of nanofluids in macroscale channels and microscale channel heat sinks.

Xue et al. [41] studied the thermal performance of a CNTs–water nanofluid in a closed two-phase thermosyphon and found that the nanofluid deteriorated the heat transfer performance. Liu et al. [68] reported that the boiling heat transfer in their thermosyphon was greatly enhanced using a Cu–water nanofluid in a miniature thermosyphon as indicated in Figure 1.14.

Ma et al. [69] reported that the heat transport capacity of an oscillating heat pipe was significantly increased using a diamond nanoparticles–water nanofluid. Liu et al. [70] conducted studies in a flat heat pipe evaporator and found that the heat transfer coefficient and CHFs of CuO–water nanofluids were enhanced by about 25% and 50%, respectively, at atmospheric pressure, whereas about 100% and 150%, respectively, at a pressure of 7.4 kPa. They also found that there was an optimum mass concentration for attaining a maximum heat transfer enhancement. Furthermore, Liu and Qiu [71] studied boiling heat transfer and the CHF of jet impingement with CuO–water nanofluids on a large flat surface and found that the boiling heat transfer was poorer while CHF was enhanced compared to that of pure water.

Flow boiling compared to pool boiling can greatly enhance the cooling performance of a microchannel heat sink by increasing the heat transfer coefficient. Furthermore, since flow boiling relies to a great degree on latent heat transfer, better temperature axial uniformity is realized both in the coolant and the wall compared to a single-phase heat sink. The question here is whether nanoparticles could further enhance an already superior performance. Lee and Mudawar [72] conducted flow boiling experiments in a microchannel heat sink using pure water and a 1% Al_2O_3 nanofluid solution. They suggested that nanofluids should not be used in microchannels due to the deposition of the nanoparticles on the channel surface. In fact, no heat transfer data were presented in their paper. Park et al. [73] studied flow boiling of nanofluids in a horizontal plain tube with an inside diameter of 8 mm, and a noticeable decrease in the heat transfer coefficient was observed and a liquid film of high particle concentration was formed on the tube surface.

In recent years, several researchers have shown the heat transfer and CHF enhancement in flow boiling with nanofluids [73–92]. A few studies relevant to the proposed research topics, that is, on the flow boiling of nanofluids in microscale channels, are available only since the last 3 years [78,79,87,89] but show enhanced heat transfer behavior. The very limited studies are not sufficient to understand the fundamentals and mechanisms. No relevant research, such as the influence on flow patterns and their transitions, pressure drops, and CHF in microscale channels, is available so far. Furthermore, investigation on the nanoparticle size's effect on flow boiling in microscale channels is not yet available. Apparently, the corresponding mechanisms and theoretical modeling are not available either. Therefore, more experiments should be conducted and new theoretical study is needed as well to explain and predict the results. Furthermore, understanding the mechanisms of flow boiling with nanofluids in microscale channels is necessary but not yet investigated so far. Especially, one could also note that some nanofluids coat the heat transfer surface, and hence, this may significantly influence the results. The surface effects need to be clearly separated from the fluidic effects in order to deduce the actual trends in the nanofluid data and thus build new models. So far, no systematic knowledge on the nanoparticle size's effects on two-phase flow patterns, pressure drop, heat transfer, and CHF in flow boiling of nanofluids and the coated surface's effect on flow boiling in channels is available. Thus, future research should be aimed at developing new fabrication technology for stable nanofluids at first, characterizing the nanofluids, modeling their physical properties, and conducting experimental and theoretical investigation on flow boiling of nanofluids in single microscale channels with various nanoparticle sizes. Surface coat effect will also be considered in the modeling aspect. Furthermore, based on the planned experimental results, new theoretical work is planned to achieve an advanced knowledge in modeling the properties of nanofluids and the evaporation of nanofluids in both macroscale and microscale channels.

1.4 Boiling Heat Transfer and CHF Mechanisms of Nanofluids

Several studies have tried to explore the mechanisms of deterioration or enhancement of nucleate pool boiling heat transfer with nanofluids. These include the decreasing of active nucleation sites from nanoparticle sedimentation on the boiling surface [45], the change of wettability of the surface [49], and nanoparticle coatings on the surface [50]. Furthermore, bubble dynamics has also been studied. Bang et al. [93] conducted visualization on nucleate pool boiling and the liquid film separating a vapor bubble from a heated surface, which was used to explain the deterioration of nucleate boiling heat transfer. You et al. [53] observed that the average size of departing bubbles increased significantly and the bubble frequency decreased significantly in nanofluids compared to those in pure water. Tu et al. [57] reported that there were smaller bubbles with no obvious changes of bubble departure frequency compared to pure water. The different bubble behaviors observed thus apparently account for the deterioration or enhancement of nucleate boiling heat transfer. However, the various contradictory results make it difficult to explain the phenomena utilizing methods for pure fluids.

Chon et al. [94] reported evaporation and dryout of nanofluid droplets on a heated surface to understand the mechanisms. They experimentally studied the thermal characteristics of evaporating nanofluid droplets using a microheater array of 32 line elements that are 100 µm wide, 0.5 µm thick, and 1.5 cm long under a constant-voltage mode. Four different nanofluids containing 2 nm Au, 30 nm CuO, 11 nm Al_2O_3, and 47 nm Al_2O_3 nanoparticles have been tested, each as 5 µL droplets with 0.5 vol.% in water. Strongly pinned nanofluid droplets are considered for a sequential evaporation process of (1) pinning, (2) liquid dominant evaporation, (3) depinning, (4) dryout, and (5) formation of a nanoparticle stain. Upon completion of the evaporation process, ring-shaped nanoparticle stains are left, the pattern of which strongly depends upon the nanoparticle sizes. Smaller nanoparticles result in relatively wider edge accumulation and more uniform central deposition, whereas larger nanoparticles make narrower and more distinctive stains at the edge with less central deposition. According to their results, nanofluid evaporation consists of three periods. First, liquid dominant evaporation occurs with steady thermal properties that are nearly identical to those of pure water with little effect of suspended nanoparticles on the overall heat and mass transfer. Next, the dryout progress characterizes the later part of evaporation, when the nanoparticle effect dominates and water level recedes. This period shows a discontinuous surge of temperature and heat flux due to the high thermal conductivity of nanoparticles, which in turn rapidly recovers to the dry heater condition while the recovery process for a pure water droplet is gradual and continual. Finally, the formation of nanoparticle stain period occurs, which strongly depends on nanoparticle size. The research focused on evaporation of droplets of nanofluids but provides some understanding of the evaporation mechanisms, which may help to explain their effect on the boiling process and two-phase phenomena.

In addition, Sefiane [95] presented a review to theoretically explore the influence of the disjoining pressure on nucleate boiling heat transfer. The disjoining pressure could push the liquid–vapor meniscus toward the vapor phase and increase the volume of the microlayer, which is similar to an increased wettability effect. Such an effect was hypothesized to increase heat transfer.

All the available studies seem to clearly conclude that the primary reason of CHF enhancement in pool boiling of nanofluids was the change of surface microstructure of the boiling surface due to a nanoparticle layer coating formed on the surface during pool boiling of nanofluids [53,54,61–67,93,96]. If this is the case, it would be easier to use an enhanced surface with a porous coating, from a practical viewpoint. Ujereh et al. [97] investigated the effects of coating silicon and copper substrates with nanotubes on nucleate pool boiling heat transfer. Different CNT array densities and area coverage were tested with FC-72. It was found that fully coating the substrate surface with CNTs was highly effective in reducing the incipience superheat and greatly enhances both nucleate boiling heat transfer coefficients and CHF. Ahn et al. [98] investigated nucleate pool boiling of refrigerant PF-5060 on two silicon wafer substrates coated with vertically aligned MWCNT *forests* of 9 μm (type-A) and 25 μm (type-B) height. It showed that the MWCNT forests enhanced CHF by 25%–28% compared to that of bare silicon (without MWCNT coating). Enhancement of nucleate boiling was not found to be sensitive to the height of MWCNT forests. In contrast, for the film boiling regime, Type-B MWCNT yielded 57% higher heat transfer at the Leidenfrost point compared to that of bare silicon. However, for the type-A MWCNT, the film boiling heat transfer values were nearly identical to the values obtained on bare silicon. SEM images with a top view of the MWCNT structures obtained before and after the experiments did not show any change in the inherent morphology of the MWCNT structures.

But further studies are still needed to clarify the heat transfer enhancement and CHF mechanism.

Examining widely quoted correlations for nucleate pool boiling heat transfer, it is not evident as to how a nanofluid will have an influence. For example, the Cooper [47] correlation Equation 1.1 is based on the reduced pressure p_r but nothing is known about the effect of nanofluids on the critical pressure or vapor pressure curve:

$$h_{nb} = 55 p_r^{0.12-0.2\log_{10} R_p} \left(-\log_{10} p_r\right)^{-0.55} M^{-0.5} q^{0.67} C \tag{1.1}$$

where
h_{nb} is the nucleate boiling heat transfer coefficient
R_p is the surface roughness (µm)
M is the molecular weight
q is the heat flux
C is a constant, which is 1 for horizontal plane surfaces and 1.7 for horizontal copper tubes according to Cooper's original paper

However, comparison with experimental data suggests that better agreement is achieved if a value of 1 is used also for horizontal tubes. Note that the heat transfer coefficient is a fairly weak function of the surface roughness parameter R_p, which is seldom well known. A value of $R_p = 1$ is suggested for technically smooth surfaces. Thus, a nanocoating may have an effect but would be very small.

Taking the Forster and Zuber [99] correlation

$$h_{nb} = 0.00122 \left[\frac{k_L^{0.79} c_{pL}^{0.45} \rho_L^{0.49}}{\sigma^{0.5} \mu_L^{0.29} h_{LV}^{0.24} \rho_V^{0.24}}\right] \Delta T_{sat}^{0.24} \Delta p_{sat}^{0.75} \tag{1.2}$$

it would predict an increase in heat transfer coefficients through the increase in liquid thermal conductivity and a decrease in heat transfer coefficients by the increase in liquid viscosity and surface tension. In Equation 1.2, k_L is the liquid thermal conductivity, c_{pL} is the liquid specific heat, ρ_L and ρ_V are liquid and vapor density, σ is the surface tension, μ_L is the liquid dynamic viscosity, h_{LV} is the latent heat, and ΔT_{sat} and Δp_{sat} are the superheated temperature and pressure.

Taking the Stephan and Abdelsalam [100] correlation for water derived by multiple regression

$$h_{nb} = 0.0546 \left[\left(\frac{\rho_V}{\rho_L}\right)^{1/2} \left(\frac{q D_{bub}}{k_L T_{sat}}\right)\right]^{0.67} \left(\frac{h_{LV} D_{bub}^2}{a_L^2}\right)^{0.248} \left(1 - \frac{\rho_V}{\rho_L}\right)^{-4.33} \frac{k_L}{D_{bub}} \tag{1.3}$$

$$D_{bub} = 0.0146\beta \left[\frac{2\sigma}{g(\rho_L - \rho_V)} \right]^{1/2} \tag{1.4}$$

Here

D_{bub} is the bubble departure diameter

The contact angle β is assigned a fixed value of 35° irrespective of the fluid

T_{sat} is the saturation temperature of the fluid in K

a_L is the liquid thermal diffusivity

g is the gravity constant

It can be summarized that the dependency of heat transfer on the liquid thermal conductivity, density, and viscosity is as follows:

$$h_{nb} \propto k_L^{-0.166}$$

$$h_{nb} \propto \sigma^{0.083}$$

Thus, it would predict a decrease in heat transfer coefficients through the increase in liquid thermal conductivity and an increase in heat transfer coefficients by the increase in surface tension while no liquid viscosity effect is concerned.

On the other hand, neither liquid thermal conductivity nor liquid viscosity is found in the CHF model of the Lienhard and Dhir [101] for pool boiling:

$$q_{crit} = 0.149 h_{LV} \rho_V \left[\frac{\sigma g (\rho_L - \rho_V)}{\rho_V^2} \right]^{1/4} \tag{1.5}$$

where q_{crit} is CHF. According to this correlation, CHF increases with increasing surface tension and liquid density. On the other hand, q_{crit} is only proportional to $\sigma^{1/4}$, so its effect is rather weak.

With respect to flow boiling heat transfer models, the nanofluid effect on the nucleate boiling contribution would be the same as in the previous section, utilizing the convective heat transfer correlation for annular flow of Kattan et al. [102]:

$$h_{cb} = 0.0133 \, Re_L^{0.69} \, Pr_L^{0.4} \frac{k_L}{\delta} \tag{1.6}$$

$$Re_L = \frac{4\rho_L u_L \delta}{\mu_L} \tag{1.7}$$

$$\Pr_L = \frac{c_{pL}\mu_L}{k_L} \tag{1.8}$$

where

h_{cb} is convective heat transfer coefficient
Re_L is the liquid film Reynolds number
$\Pr_L$ is the liquid Prandtl number
δ is the liquid film thickness

It can be summarized that the dependency of heat transfer on the liquid thermal conductivity, density, and viscosity is as follows:

$$h_{cb} \propto k_L^{0.6}$$

$$h_{cb} \propto \mu_L^{-0.29}$$

Thus, this predicts an increase in heat transfer coefficient through the increase in the liquid thermal conductivity but a decrease in heat transfer coefficient by the increase in liquid viscosity, while no surface tension effect is concerned.

Regarding the CHF in saturated flow boiling in microchannels, the recent empirical correlation of Wojtan et al. [103]

$$q_{crit} = 0.437 \left(\frac{\rho_V}{\rho_L}\right)^{0.073} \text{We}_L^{-0.24} \left(\frac{L_H}{D}\right)^{-0.72} G h_{LV} \tag{1.9}$$

$$\text{We}_L = \frac{G^2 L_H}{\rho_L \sigma} \tag{1.10}$$

can be used for analysis here, where We_L is the Weber number based on heated length, D is the tube diameter, L_H is the heated length, and G is the mass flux. Similar to the CHF model of Lienhard and Dhir [101] for pool boiling, neither liquid thermal conductivity nor liquid viscosity is found in this expression. However, CHF increases with increasing surface tension and liquid density according to this expression.

From the preceding analysis, it is clearly shown that the physical properties such as surface tension, liquid density, and viscosity have an effect on nucleate pool boiling heat transfer, convective flow boiling, and CHF in both pool and flow boiling processes. So far, the lack of knowledge of these physical properties of nanofluids greatly limits an evaluation of the possible effect. This also poses a serious question: Which physical properties should we use to reduce experimental data for nanofluids? The data reduction methods used might be one of the reasons why the available experiments are contradictory.

Furthermore, nucleation density site, bubble dynamics, thin film evaporation, dryout, liquid–vapor interfacial force, and boiling surface structures are the main factors that affect nucleate boiling heat transfer and CHF. For nanofluids, the size and type of nanoparticles could be important but it is still unclear they would affect the underlying mechanisms. Considering the controversies from the previous studies, the aggregation of nanofluids could be an important factor affecting boiling performance, which needs to be clarified quantitatively. Furthermore, the mechanisms that explain the substantial increase in CHF still need to be verified.

1.5 Challenges and Future Research Needs

There are many challenges in the study of nanofluid two-phase flow and thermal physics. The available experimental data on nucleate pool boiling heat transfer of nanofluids are quite limited, and many conflicts exist between different studies on the heat transfer characteristics so far. The inconsistencies indicate that the understanding of the thermal behaviors of nanofluids related to nucleate pool boiling heat transfer is still poor. The results on CHF enhancement by nanofluids are consistent with each other; however, the mechanism responsible for this is not yet clear. Advanced physical models are required to explain and predict the influence of nanoparticles on nucleate pool boiling and CHF.

Explanations and new theories are also needed to take into account all the important characteristics of nanofluids on flow boiling. So far, no systematic knowledge of their effects is available. Considering the fact that CNTs can enhance both nucleate pool boiling and CHF of R123 and R134a [80,81], it is recommended that both experimental and theoretical investigations on flow boiling of CNT nanofluids (such as R134a with CNTs) in single or multimicrochannels should be performed. Flow boiling with other nanofluids should also be investigated in the future. Based on the experimental results, new theoretical work should be developed to achieve an advanced knowledge in the modeling of the evaporation of nanofluids.

Flow boiling compared to nucleate pool boiling can greatly enhance the cooling performance of a microchannel heat sink by increasing the heat transfer coefficient. Furthermore, since flow boiling relies to a great degree on latent heat transfer, better temperature axial uniformity is realized both in the coolant and the wall compared to a single-phase heat sink. The question here is whether nanoparticles could further enhance an already superior heat transfer performance. Furthermore, it would be very valuable indeed to see if the increase in nucleate pool boiling CHF also occurs in flow boiling CHF, for which there may be numerous high heat dissipation applications in microdevices and the new generation of CPU chips, just to name a few. On the other hand, it must be pointed out that nanoparticles should only be used for processes whose exit quality is less than that of the onset of dryout, in order to minimize the deposition of the nanoparticles on the channel wall.

Furthermore, experimental and theoretical studies on the condensation of nanofluids inside macro- and microchannels should be systematically conducted as well. Other two-phase flow characteristics such as flow patterns and two-phase pressure drops should be investigated. The flow patterns should be related to the corresponding heat transfer and pressure drop characteristics. Flow pattern–based heat transfer and pressure drop prediction methods should be also developed based on a wide range of experimental data.

Apparently, there are many challenges in nanofluid two-phase flow and thermal physics, which is a new interdisciplinary research frontier of nanotechnology. Much work is needed to achieve the fundamental knowledge and practical applications.

1.6 Concluding Remarks

The research of boiling and CHF phenomena of nanofluids is still in its infancy. Many controversies exist with numerous conflicting experimental results and trends. Through this comprehensive literature review, the following future research needs have been identified:

1. Physical properties such as liquid density and viscosity, surface tension, and specific heat have a significant effect on nanofluid two-phase flow and thermal physics such as nucleate pool boiling, convective flow boiling, and CHF in both pool and flow boiling processes. To properly present the experimental results and to understand the physical mechanisms related to the two-phase and thermal phenomena, the nanofluid physical properties should be systematically investigated to set up a consistent database of physical properties in addition to thermal conductivity.

2. Nucleate pool boiling heat transfer and its mechanisms should be further investigated. The inconsistencies between different studies should be clarified. Furthermore, the effect of nanoparticle size and type on heat transfer should be studied. The heat transfer mechanisms responsible for these trends should be identified, and explanation should be provided as to why nucleate heat transfer may be enhanced or decreased. Data should also be segregated by fluids that deposit on the boiling surface and those that do not, in order to prove if the fluid alone can enhance performance.

3. CHF in pool boiling should be systematically investigated and the mechanisms responsible for its delay to higher heat fluxes definitively identified. Furthermore, a new model for CHF should be developed based on the experimental nanofluid data and the CHF mechanisms.

4. More experiments on nanofluid two-phase flow and flow boiling should be conducted in both macro- and microchannels to evaluate the potential benefits of nanofluids. These should also include heat transfer performance, CHF, two-phase flow patterns, and pressure drop in various types of channels. Especially, the two-phase flow and heat transfer characteristics should be related to the corresponding flow patterns.

5. The sediment or coating of nanoparticles on the heat transfer surface is a big question that needs to be resolved. For example, if such a coating is beneficial, then it could be applied more easily using a coating process rather than nanofluid deposition. If such a nanoparticle layer has adverse effects, then ways to prevent it are needed or the correct nanofluids should be found.

6. Prediction methods that include nanofluid effects should be developed.

7. The application of nanofluid two-phase flow and thermal physics in real world should also be targeted in order to get more appropriate experience.

8. There are still too many unresolved problems with respect to our knowledge of two-phase nanofluids, but great efforts should be made to contribute to this new research frontier in both fundamental and applied research.

References

1. S.U.S. Choi, Enhancing thermal conductivity of fluids with nanoparticles, in *Developments and Applications of Non-Newtonian Flows*, FED-vol. 231/MD-vol. 66, ASME pp. 99–105, 1995.

2. X.Q. Wang, A.S. Majumdar, Heat transfer characteristics of nanofluids: A review, *Int. J. Therm. Sci.* 46 (2007): 1–19.

3. L. Cheng, L. Liu, Boiling and two phase flow phenomena of refrigerant-based nanofluids: Fundamentals, applications and challenges, *Int. J. Refrig.* 36 (2013): 421–446.

4. L. Cheng, E.P. Bandarra Filho, J.R. Thome, Nanofluid two-phase flow and thermal physics: A new research frontier of nanotechnology and its challenges, *J. Nanosci. Nanotechnol.* 8 (2008): 3315–3332.

5. S.K. Das, S.U.S. Choi, H. Patel, Heat transfer in nanofluids—A review, *Heat Transfer Eng.* 27(10) (2006): 3–19.

6. L. Cheng, Nanofluid heat transfer technologies, *Recent Pat. Eng.* 3(1) (2009): 1–7.

7. L. Cheng, Fundamental issues of critical heat flux phenomena during flow boiling in microscale-channels and nucleate pool boiling in confined spaces, *Heat Transfer Eng.* 34 (2013): 1011–1043.

8. J.R. Thome, Boiling in microchannels: A review of experiment and theory, *Int. J. Heat Fluid Flow* 25 (2004): 128–139.

9. L. Cheng, D. Mewes, Review of two-phase flow and flow boiling of mixtures in small and mini channels, *Int. J. Multiphase Flow* 32 (2006): 183–207.

10. S.G. Kandlikar, Fundamental issues related to flow boiling in minichannels and microchannels, *Exp. Therm. Fluid Sci.* 26 (2002): 38–47.

11. S.S. Mehendal, A.M. Jacob, R.K. Shah, Fluid flow and heat transfer at micro- and meso-scales with applications to heat exchanger design, *Appl. Mech. Rev.* 53 (2000): 175–193.

12. J.R. Thome, The new frontier in heat transfer: Microscale and nanoscale technologies, *Heat Transfer Eng.* 27(9) (2006): 1–3.

13. J.R. Thome, *Enhanced Boiling Heat Transfer*, Hemisphere Publishing Corporation, New York, 1990.

14. L. Cheng, D. Mewes, A. Luke, Boiling phenomena with surfactants and polymeric additives: A state-of-the-art review, *Int. J. Heat Mass Transfer* 50 (2007): 2744–2771.

15. J.A. Eastman, S.U.S. Choi, S. Li, W. Yu, L.J. Thompson, Anomalously increased effective thermal conductivities of ethylene glycol-based nanofluids containing copper nanoparticles, *Appl. Phys. Lett.* 78 (2001): 718–720.

16. Y. Xuan, Q. Li, Heat transfer enhancement of nanofluids, *Int. J. Heat Fluid Flow* 21 (2000): 58–64.

17. S.K. Das, N. Putra, P. Thiesen, W. Roetzel, Temperature dependence of thermal conductivity enhancement for nanofluids, *J. Heat Transfer* 125 (2003): 567–574.

18. S. Lee, S.U.S. Choi, S. Li, J.A. Eastman, Measuring thermal conductivity of fluids containing oxide nanoparticles, *J. Heat Transfer* 121 (1999): 280–289.

19. S.P. Jang, S.U.S. Choi, Role of Brownian motion in the enhanced thermal conductivity of nanofluids, *Appl. Phys. Lett.* 84 (2004): 219–246.

20. T.K. Hong, H.-S. Yang, C.J. Choi, Study of the enhanced thermal conductivity of Fe nanofluids, *J. Appl. Phys.* 97(6) (2005): 1–4.

21. S.P. Jang, S.U.S. Choi, Effects of various parameters on nanofluid thermal conductivity, *J. Heat Transfer* 129 (2007): 617–623.

22. S.M.S. Murshed, K.C. Leong, C. Yang, Enhanced thermal conductivity of TiO_2–water based nanofluids, *Int. J. Therm. Sci.* 44 (2005): 367–373.

23. S.M.S. Murshed, K.C. Leong, C. Yang, Investigations of thermal conductivity and viscosity of nanofluids, *Int. J. Therm. Sci.* 47 (2008): 560–568.

24. J.Y. Huang, S. Chen, Z.Q. Wang, K. Kempa, Y.M. Wang, S.H. Jo, G. Chen, M.S. Dresselhaus, Z.F. Ren, Superplastic single-walled carbon nanotubes, *Nature* 439 (2006): 281.

25. M.J. Biercuk, M.C. Llaguno, M. Radosavljevic, J.K. Hyun, A.T. Johnson, Carbon nanotube composites for thermal management, *Appl. Phys. Lett.* 80 (2002): 2767–2769.

26. J. Hone, M. Whitney, A. Zettl, Thermal conductivity of single-walled carbon nanotubes, *Synth. Met.* 103 (1999): 2498–2499.

27. S. Berber, Y.K. Kwon, D. Tomanek, Unusually high thermal conductivity of carbon nanotubes, *Phys. Rev. Lett.* 84 (2000): 4613–4616.

28. P. Kim, L. Shi, A. Majumdar, P.L. Mceuen, Thermal transport measurements of individual multiwalled nanotubes, *Phys. Rev. Lett.* 87 (2001): 215502.

29. S.U.S. Choi, Z.G. Zhang, W. Yu, F.E. Lockwood, E.A. Grulke, Anomalous thermal conductivity enhancement in nanotube suspensions, *Appl. Phys. Lett.* 79 (2001): 2252–2254.

30. H. Xie, H. Lee, W. Youn, M. Choi, Nanofluids containing multiwalled carbon nanotubes and their enhanced thermal conductivities, *J. Appl. Phys.* 94 (2003): 4967–4971.

31. M.S. Liu, M.C.C. Lin, I.T. Huang, C.C. Wang, Enhancement of thermal conductivity with carbon nanotube for nanofluids, *Int. Commun. Heat Mass Transfer* 32 (2005): 1202–1210.

32. M.J. Assael, I.N. Metaxa, J. Arvanitidis, D. Christofilos, C. Lioutas, Thermal conductivity enhancement in aqueous suspensions of carbon multi-walled and double-walled nanotubes in the presence of two different dispersants, *Int. J. Thermophys.* 26 (2005): 647–664.

33. D.S. Wen, Y.L. Ding, Effective thermal conductivity of aqueous suspensions of carbon nanotubes (carbon nanotubes nanofluids), *J. Thermophys. Heat Transfer* 18 (2004): 481–485.

34. Y. Ding, H. Alias, D. Wen, R.A. Williams, Heat transfer of aqueous suspensions of carbon nanotubes, *Int. J. Heat Mass Transfer* 49 (2005): 240–250.

35. R.L. Hamilton, O.K. Crosser, Thermal conductivity of heterogeneous two-component systems, *Ind. Eng. Chem. Fund.* 1(3) (1962): 187–191.

36. P. Keblinski, S.R. Phillpot, S.U.S. Choi, J.A. Eastman, Mechanisms of heat flow in suspensions of nano-sized particles (nanofluids), *Int. J. Heat Mass Transfer* 45 (2002): 855–863.

37. Y. Xuan, Q. Li, W. Hu, Aggregation structure and thermal conductivity of nanofluids, *AIChE J.* 49 (2003): 1038–1043.

38. D.P. Kulkarni, D.K. Das, S.L. Patil, Effect of temperature on rheological properties of copper oxide nanoparticles dispersed in propylene glycol and water mixture, *J. Nanosci. Nanotechnol.* 7 (2007): 2318–2322.

39. D.P. Kulkarni, D.K. Das, G.A. Chukwu, Temperature dependent rheological property of copper oxide nanoparticles suspension (nanofluid), *J. Nanosci. Nanotechnol.* 6 (2006): 1150–1154.

40. C.T. Nguyen, F. Desgranges, N. Galanis, G. Roy, T. Maré, S. Boucher, H.A. Minsta, Viscosity data for Al_2O_3–water nanofluid—Hysteresis: Is heat transfer enhancement using nanofluids reliable? *Int. J. Therm. Sci.* 47 (2008): 103–111.

41. H.S. Xue, J.R. Fan, Y.C. Hu, R.H. Hong, K.F. Cen, The interface effect of carbon nanotube suspension on thermal performance of a two-phase closed thermosyphon, *J. Appl. Phys.* 100 (2006): 104909.

42. D. Shin, D. Banerjee, Enhanced specific heat of silica nanofluid, *ASME J. Heat Transfer* 133 (2010): 024510.

43. D. Shin, D. Banerjee, Enhancement of specific heat capacity of high-temperature silica-nanofluids synthesized in alkali chloride salt eutectics for solar thermal-energy storage applications, *Int. J. Heat Mass Transfer* 54 (2011): 1064–1070.

44. H. Tiznobaik, D. Shin, Enhanced specific heat capacity of high-temperature molten salt-based nanofluids, *Int. J. Heat Mass Transfer* 54 (2011): 1064–1070.

45. S. Witharana, Boiling of refrigerants on enhanced surfaces and boiling of nanofluids, PhD thesis, The Royal Institute of Technology, Stockholm, Sweden, 2003.

46. C.Y. Yang, D.W. Liu, Effect of nano-particles for pool boiling heat transfer of refrigerant 141B on horizontal tubes, *Int. J. Microscale Nanoscale Therm. Fluid Transp. Phenom.* 1(3) (2010): 233–243.

47. M.G. Cooper, Saturation nucleate pool boiling: A simple correlation, *Int. Chem. Eng. Symp. Ser.* 86 (1984): 785–792.

48. D. Wen, Y. Ding, Experimental investigation into pool boiling heat transfer of aqueous based γ-alumina nanofluids, *J. Nanoparticle Res.* 7 (2005): 265–274.

49. S.K. Das, N. Putra, W. Roetzel, Pool boiling characteristics of nanofluids, *Int. J. Heat Mass Transfer* 46 (2003): 851–862.

50. I.C. Bang, S.H. Chang, Boiling heat transfer performance and phenomena of Al_2O_3–water nano-fluids from a plain surface in a pool, *Int. J. Heat Mass Transfer* 48 (2005): 2407–2419.

51. C.H. Li, B.X. Wang, X.F. Peng, Experimental investigations on boiling of nanoparticle suspensions, in: *2003 Boiling Heat Transfer Conference*, Montego Bay, Jamaica, 2003.

52. S.K. Das, N. Putra, W. Roetzel, Pool boiling characteristics of nanofluids on horizontal narrow tubes, *Int. J. Multiphase Flow* 29 (2003): 1237–1247.

53. S.M. You, J.H. Kim, K.H. Kim, Effect of nanoparticles on critical heat flux of water in pool boiling heat transfer, *Appl. Phys. Lett.* 83 (2003): 3374–3376.

54. J.H. Kim, K.H. Kim, S.M. You, Pool boiling heat transfer in saturated nanofluids, in: *2004 ASME International Mechanical Engineering Congress and Exposition*, IMECE2004-61108, November 13–30, 2004, Anaheim, CA, pp. 621–628, *Heat Transfer*, Vol. 2, 2004.

55. P. Vassallo, R. Kumar, S. D'Amico, Pool boiling heat transfer experiments in silica-water nano-fluids, *Int. J. Heat Mass Transfer* 47 (2004): 407–411.

56. G. Prakash Narayan, A.K.B.G. Sateesh, S.K. Das, Effect of surface orientation on pool boiling heat transfer of nanoparticle suspensions, *Int. J. Multiphase Flow* 34 (2008): 145–160.

57. J.P. Tu, N. Dinh, T. Theofanous, An experimental study of nanofluid boiling heat transfer, in: *Proceedings of the Sixth International Symposium on Heat Transfer*, Beijing, China, 2004.

58. D.S. Wen, Y.L. Ding, R.A. Williams, Pool boiling heat transfer of aqueous based TiO_2 nanofluids, *J. Enhanc. Heat Transfer* 13 (2006): 231–244.

59. M. Chopkar, A.K. Das, I. Manna, P.K. Das, Pool boiling heat transfer characteristics of ZrO_2–water nanofluids from a flat surface in a pool, *Heat Mass Transfer*, 44 (2008): 999–1004.

60. K.J. Park, D. Jung, Boiling heat transfer enhancement with carbon nanotubes for refrigerants used in building air-conditioning, *Energy Build.* 39 (2007): 1061–1064.

61. S.J. Kim, I.C. Bang, J. Buongiorno, L.W. Hu, Surface wettability change during pool boiling of nanofluids and its effect on critical heat flux, *Int. J. Heat Mass Transfer* 50 (2007): 4105–4116.

62. S.J. Kim, I.C. Bang, J. Buongiorno, L.W. Hu, Study of pool boiling and critical heat flux enhancement in nanofluids, *Bull. Polish Acad. Sci.* 55(2) (2007): 211–216.

63. H. Kim, J. Kim, M.H. Kim, Effect of nanoparticles on CHF enhancement in pool boiling of nano-fluids, *Int. J. Heat Mass Transfer* 49 (2006): 5070–5074.
64. H.D. Kim, J. Kim, M.H. Kim, Experimental studies on CHF characteristics of nano-fluids at pool boiling, *Int. J. Multiphase Flow* 33 (2007): 691–706.
65. H.D. Kim and M.H. Kim, Critical heat flux behavior in pool boiling of water–TiO_2 nano-fluids, *Appl. Phys. Lett.* 91 (2007): 014104.
66. D. Milanova, R. Kumar, Role of ions in pool boiling heat transfer of pure and silica nanofluids, *Appl. Phys. Lett.* 87 (2005): 233107.
67. H.S. Xue, J.R. Fan, R.H. Hong, Y.C. Hu, Characteristic boiling curve of carbon nanotube nanofluid as determined by the transient calorimeter technique, *Appl. Phys. Lett.* 90 (2007): 184107.
68. Z.H. Liu, X.F. Yang, G.L. Guo, Effect of nanoparticles in nanofluids on thermal performance in a miniature thermosyphon, *J. Appl. Phys.* 102 (2007): 013526.
69. H.B. Ma, U.S. Choi, M. Tirumala, C. Wilson, Q. Yu, K. Park, An experimental investigation of heat transport capability in a nanofluid oscillating heat pipe, *J. Heat Transfer* 128 (2006): 1213–1216.
70. Z.H. Liu, J.G. Xiong, R. Bao, Boiling heat transfer characteristics of nanofluids in a flat heat pipe evaporator with micro-grooved heating surface, *Int. J. Multiphase Flow* 33 (2007): 1284–1295.
71. Z.H. Liu, Y.H. Qiu, Boiling heat transfer characteristics of nanofluids jet impingement on a plate surface, *Heat Mass Transfer* 43 (2007): 699–706.
72. J. Lee, I. Mudawar, Assessment of the effectiveness of nanofluids for single-phase and two-phase heat transfer in micro-channels, *Int. J. Heat Mass Transfer* 50 (2007): 452–463.
73. Y. Park, A. Sommers, L. Liu, G. Michna, A. Joardar, A. Jacobi, Nanoparticles to enhance evaporative heat transfer, in: *The 22nd International Congress of Refrigeration*, August 21–26, 2007, Beijing, China, in CD-Rom, Paper number: ICR07-B1-309.
74. K. Henderson, Y.-G. Park, L. Liu, A.M. Jacobi, Flow boiling heat transfer of R-134a-based nanofluids in a horizontal tube, *Int. J. Heat Mass Transfer* 53 (2010): 944–951.
75. S.J. Kim, T. McKrell, J. Buongiorno, L.W. Hu, Experimental study of flow critical heat flux in alumina-water, zinc-oxide-water, and diamond-water nanofluids, *ASME J. Heat Transfer* 131(4) (2009): 043204/1–043204/7.
76. K.B. Rana, A.K. Rajvanshi, G.D. Agrawal, A visualization study of flow boiling heat transfer with nanofluids, *J. Vis.* 16 (2013): 133–143.
77. T.I. Kim, W.J. Chang, S.H. Chang, Flow boiling CHF enhancement using Al_2O_3 nanofluid and an Al_2O_3 nanoparticle deposited tube, *Int. J. Heat Mass Transfer* 54 (2011): 2021–2025.
78. S. Vafaei, D. Wen, Critical heat flux (CHF) of subcooled flow boiling of alumina nanofluids in a horizontal microchannel, *ASME J. Heat Transfer* 132 (2010): 1–7.
79. S. Vafaei, D. Wen, Flow boiling heat transfer of alumina nanofluids in single microchannels and the roles of nanoparticles, *J. Nanoparticle Res.* 13 (2011): 1063–1073.
80. H.S. Ahn, M.H. Kim, The effect of micro/nanoscale structures on CHF enhancement, *Nucl. Eng. Technol.* 43 (2011): 205–216.
81. H.S. Ahn, S.H. Kang, M.H. Kim, Visualized effect of alumina nanoparticles surface deposition on water flow boiling heat transfer, *Exp. Therm. Fluid Sci.* 37 (2012): 154–163.

82. S.W. Lee, S.D. Park, S. Kang, S.M. Kim, H. Seo, D.W. Lee, I.C. Bang, Critical heat flux enhancement in flow boiling of Al_2O_3 and SiC nanofluids under low pressure and low flow conditions, *Nucl. Eng. Technol.* 44 (2012): 429–436.

83. X.F. Yang, Z.H. Liu, Flow boiling heat transfer in the evaporator of a loop thermosyphon operating with CuO based aqueous nanofluid, *Int. J. Heat Mass Transfer* 55 (2012): 7375–7384.

84. S.J. Kim, T. McKrell, J. Buongiorno, L.W. Hu, Subcooled flow boiling heat transfer of dilute alumina, zinc oxide, and diamond nanofluids at atmospheric pressure, *Nucl. Eng. Des.* 240 (2010): 1186–1194.

85. H.S. Ahn, H. Kim, H. Jo, S. Kang, W. Chang, M.H. Kim, Experimental study of critical heat flux enhancement during forced convective flow boiling of nanofluid on a short heated surface, *Int. J. Multiphase Flow* 36 (2010): 375–384.

86. H.S. Ahn, S. Kang, H. Jo, H. Kim, M.H. Kim, Visualization study of the effects of nanoparticles surface deposition on convective flow boiling CHF from a short heated wall, *Int. J. Multiphase Flow* 37 (2011): 215–228.

87. M. Boudouh, H.L. Gualous, M. De Labachelerie, Local convective boiling heat transfer and pressure drop of nanofluid in narrow rectangular channels, *Appl. Therm. Eng.* 30 (2010): 2619–2631.

88. T. Lee, J.H. Lee, Y.H. Jeong, Flow boiling critical heat flux characteristics of magnetic nanofluid at atmospheric pressure and low mass flux conditions, *Int. J. Heat Mass Transfer* 56 (2013): 101–106.

89. A.A. Chehade, H.L. Gualous, S. Le Masson, F. Fardoun, A. Besq, Boiling local heat transfer enhancement in minichannels using nanofluids, *Nanoscale Res. Lett.* 8 (2013): 1–20.

90. G. Dewitt, T. Makrell, J. Buongiorno, L.W. Hu, R.J. Park, Experimental study of critical heat flux with alumina-water nanofluids in downward-facing channels for in-vessel retention applications, *Nucl. Eng. Technol.* 45 (2013): 335–346.

91. S.W. Lee, K.M. Kim, I.C. Bang, Study on flow boiling critical heat flux enhancement of graphene oxide/water nanofluid, *Int. J. Heat Mass Transfer* 65 (2013): 348–356.

92. Z. Abedini, A. Behzadmehr, H. Rajabnia, S.M.H. Sarvari, S.H. Mansouri, Experimental investigation and comparison of subcooled flow boiling of TiO_2 nanofluid in a vertical and horizontal tube, *Proc. Inst. Mech. Eng., C. J. Mech. Eng. Sci.* 227 (2013): 1742–1753.

93. I.C. Bang, S.H. Chang, W.P. Baek, Direct observation of a liquid film under a vapor environment in a pool boiling using a nanofluid, *Appl. Phys. Lett.* 86 (2005): 134107.

94. X.H. Chon, S. Park, J.B. Tipton Jr., K.D. Kilhm, Effect of nanoparticle sizes and number densities on the evaporation and dryout characteristics for strongly pinned nanofluid droplets, *Langmuir* 23 (2007): 2953–2960.

95. K. Sefiane, On the role of structural disjoining pressure and contact line pinning in critical heat flux enhancement during boiling of nanofluids, *Appl. Phys. Lett.* 89 (2006): 044106.

96. K.J. Park, D. Jung, S.E. Shim, Nucleate boiling heat transfer in aqueous solutions with carbon nanotubes up to critical heat fluxes, *Int. J. Multiphase Flow* 35 (2009): 525–532.

97. S. Ujereh, T. Fisher, I. Mudawar, Effects of carbon nanotube arrays on nucleate pool boiling, *Int. J. Heat Mass Transfer* 50 (2007): 4023–4038.

98. H.S. Ahn, N. Sinha, M. Zhang, D. Banerjee, S. Fang, R.H. Baughman, Pool boiling experiments on multiwalled carbon nanotube (MWCNT) forests, *ASME J. Heat Transfer*, 128 (2006): 1335–1342.

99. H.K. Forster, N. Zuber, Dynamics of vapor bubbles and boiling heat transfer, *AIChE J.* 1 (1955): 531–535.

100. K. Stephan, M. Abdelsalam, Heat transfer correlation for natural convection boiling, *Int. J. Heat Mass Transfer* 23 (1980): 73–87.

101. J.H. Lienhard, V.K. Dhir, Peak pool boiling heat-flux measurements on finite horizontal flat plates, *ASME J. Heat Transfer* 95 (1973): 477–482.

102. N. Kattan, J.R. Thome, D. Favrat, Flow boiling in horizontal tubes: Part 3. Heat transfer model based on flow pattern, *ASME J. Heat Transfer* 120 (1998): 156–165.

103. L. Wojtan, R. Revellin, J.R. Thome, Investigation of critical heat flux in single, uniformly heated microchannels, *Exp. Therm. Fluid Sci.* 30 (2006): 765–774.

2

Modeling for Heat Transfer of Nanofluids Using a Fractal Approach

Boqi Xiao

CONTENTS

2.1 Introduction 32
2.2 Fractal Analysis of Nanoparticles in Nanofluids 33
2.3 Modeling for Effective Thermal Conductivity of Nanofluids
 Using a Fractal Approach 36
    2.3.1 Fractal Model 36
    2.3.2 Results and Discussions 41
    2.3.3 Summary and Conclusions 43
2.4 Modeling for Convective Heat Transfer of Nanofluids Based on
 Fractal Theory and Monte Carlo Simulation 44
    2.4.1 Formulation of the Proposed Model 44
    2.4.2 Methodology for the Monte Carlo Technique 46
    2.4.3 Results and Discussions 48
    2.4.4 Summary and Conclusions 50
2.5 Fractal Model for Critical Heat Flux of Nanofluids 51
    2.5.1 Fractal Method and Calculations 51
    2.5.2 Results and Discussions 55
    2.5.3 Summary and Conclusions 58
2.6 Fractal Analysis of Pool Boiling Heat Transfer in Nanofluids 59
    2.6.1 Modeling Heat Flux of Subcooled Pool Boiling
 in Fractal Nanofluids 59
    2.6.2 Analytical Model for Nucleate Pool Boiling Heat Transfer
 of Nanofluids 64
    2.6.3 Results and Discussions 67
    2.6.4 Summary and Conclusions 71
Nomenclature 72
Greek Letters 73
Subindexes 74
Acknowledgments 74
References 74

2.1 Introduction

Recent advances in nanotechnology have promoted a rapid development in nanofluids, fluids suspensions of nanoparticulate solids including particles, nanofibers, and nanotubes. Fluids with suspended nanoparticles are called nanofluids. Nanofluid technology has emerged as a new heat transfer technique in recent years (Prasher et al. 2005; Shima et al. 2009; Xiao et al. 2009, 2010, 2013a,b; Xiao 2013). Common base fluids include water and organic liquids. *Nanofluids* is the term used to describe the fluids that contain colloidal dispersion of nanometer-sized solid particles. In general, nanofluids are suspensions of solid nanoparticles with sizes typically of 1–100 nm in a base fluid. Two important characteristics of these nanofluids observed to date are (1) they are better conductors of heat than the base fluid itself and (2) they are more stable compared to fluids containing microparticles, in which the particles tend to settle down after a finite period of time. Since nanoparticles offer extremely large total surface areas, nanofluids exhibit superior thermal properties relative to conventional heat transfer fluids and fluids containing micrometer-sized particles. The transport properties of nanofluids have been a topic of intense research due to their prospective technological applications in electronics cooling and heat transfer. For example, one area of interest is in boiling flow systems where the ability to enhance heat transfer would improve the overall efficiency of the systems, reduce operational cost, and provide greater safety margins if the maximum heat flux limit could be increased.

It has long been recognized that the study of basic heat transport processes of nanofluids can heighten the comprehensive understanding of physical phenomena such as convection and boiling. Because of the various speculated uses of nanofluids, it has become important to know more about the heat transfer properties of fluids. However, it is complicated to analyze the heat transfer behaviors within mechanical systems, especially when different heat transfer mechanisms are depicted together in literatures. From a mechanistic viewpoint, although the influences of some parameters such as the average diameter of nanoparticles, the volumetric nanoparticle concentration, and the thermal conductivity of nanoparticles have been discussed, an overall mechanistic description is still unavailable. In addition, each model/correlation has its disadvantages because of the limitations in experimental conditions. It is a challenge to calculate the heat transfer of nanofluids due to the extremely complicated mechanisms of heat transfer as well as the interrelationship between heat flux and the size of nanoparticles and volumetric nanoparticle concentration.

2.2 Fractal Analysis of Nanoparticles in Nanofluids

Euclidean geometry can be used to depict ordered objects such as curves, surfaces, and cubes, using integer dimensions 1, 2, and 3, respectively. However, it is found that numerous objects are disordered and irregular in nature such as coastlines, rough surfaces, mountains, rivers, lakes, and islands. They do not follow the Euclidean description since it employs scale-dependent measures of length, area, and volume. Such objects are called fractals and are described using a nonintegral dimension called the fractal dimension (Mandelbrot 1982). Two examples (Mandelbrot 1982) of well-known fractals are shown in Figure 2.1. Fractals are best constructed in a recursive way. For example, the Koch curve is constructed starting with a unit segment.

Fractals exhibit the self-similarity over a range of length scales. The fractal object is related to the length scale by a power law (Mandelbrot 1982):

$$M(\lambda) \sim \lambda^{d_f} \tag{2.1}$$

where
d_f is the fractal dimension of an object
$M(\lambda)$ can be a quantity, volume, or area of the object
λ is the length scale

The experiment combined with fractal geometry theory is applied to predict effective thermal conductivity of nanofluids (Wang et al. 2003). It has been

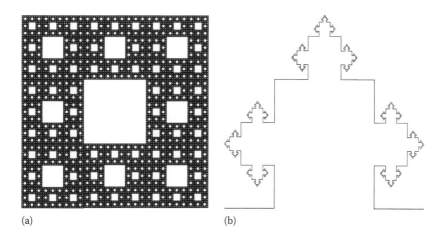

(a)   (b)

FIGURE 2.1
Examples of fractals: (a) Sierpinski carpet and (b) Koch curve.

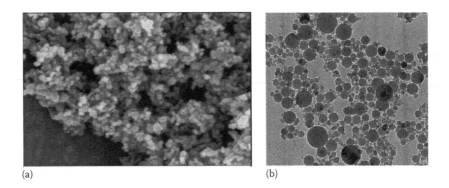

(a)   (b)

FIGURE 2.2

(a) An SEM micrograph of TiO_2 nanoparticles (From He, Y.R. et al., *Int. J. Heat Mass Transfer,* 50(11–12), 2272, 2007) and (b) TEM photograph of Al_2O_3 nanoparticles. (From Bang, I.C. and Chang, S.H., *Int. J. Heat Mass Transfer,* 48(12), 2407, 2005.)

shown that the size distribution of nanoparticles in nanofluids follows the fractal power law (Wang et al. 2003). Figure 2.2a shows a scanning electron microscopy (SEM) image of the sample for TiO_2 nanoparticles at $D_{av} = 20$ nm (average diameter of nanoparticles) (He et al. 2007). As shown in the transmission electron microscopy (TEM) image in Figure 2.2b, Al_2O_3 nanoparticles have a spherical shape (Bang and Chang 2005). The size has a normal distribution in the range of 10–100 nm ($D_{av} = 47$ nm) (Bang and Chang 2005). The analysis showed a wide distribution of particles as shown in Figure 2.3

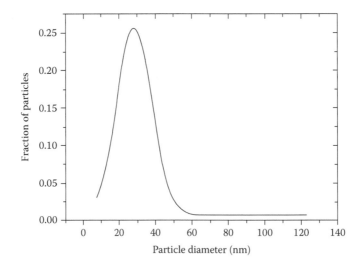

FIGURE 2.3

The size distribution of Al_2O_3 nanoparticles. (From Bang, I.C. and Chang, S.H., *Int. J. Heat Mass Transfer,* 48(12), 2407, 2005.)

(Bang and Chang 2005). From Figures 2.1 through 2.3, it can be seen that the number of nanoparticles is very large in nanofluids. Besides, they are different in size.

The fractal dimension (d_f) of nanoparticles is an important parameter for the fractal analysis of nanofluids. The relation between the fractal dimension of nanoparticles and nanoparticle concentration can be expressed as (Xiao et al. 2009, 2013a,b; Xiao 2013)

$$\phi = \beta^{2-d_f}$$

(2.2)

where
   ϕ is the nanoparticle concentration
   $D_{min}/D_{max} = \beta$, D_{max} and D_{min} are the maximum and minimum diameter of nanoparticles, respectively

The theory of fractals aids in analyzing the response of practical systems besides examining the complex geometries of nature. As universally received, the fractal technique is now used to predict the transport properties of fluids, such as permeability of porous medium (Yu and Cheng 2002a), heat transfer of fluids (Xiao et al. 2009, 2013a,b), and spontaneous imbibition in porous medium (Cai and Yu 2010, 2011; Cai et al. 2012). The nanoparticles in nanofluids have been proven to be fractal objects (Wang et al. 2003; Xiao et al. 2009, 2010, 2013a,b; Xiao 2013). It is shown that the distribution of nanoparticles in suspension exhibits fractal property. This means that the fractal theory may be used to predict the transport property of nanofluids. The cumulative number N of nanoparticles in fluids with sizes greater than or equal to D can be expressed according to the fractal scaling law as (Yu and Cheng 2002a; Xiao et al. 2009, 2013a,b)

$$N(L \geq D) = \left(\frac{D_{max}}{D}\right)^{d_f} \quad \text{with } D_{min} \leq D \leq D_{max}$$

(2.3)

Differentiating Equation 2.3 with respect to D yields

$$-dN = d_f D_{max}^{d_f} D^{-(d_f+1)} dD$$

(2.4)

Equation 2.4 gives the nanoparticles number between the nanoparticles size D and $D + dD$. The negative sign in Equation 2.4 implies that nanoparticles number decreases with the increase of nanoparticles size, and $-dN > 0$. The total number of nanoparticles in nanofluids from the minimum diameter D_{min} to the maximum diameter D_{max} can be derived from Equation 2.3 as

$$N_t(L \geq D_{min}) = \left(\frac{D_{max}}{D_{min}}\right)^{d_f}$$

(2.5)

Dividing Equation 2.4 by Equation 2.5 results in

$$-\frac{dN}{N_t} = d_f D_{\min}^{d_f} D^{-(d_f+1)} dD = f(D)dD \tag{2.6}$$

where $f(D) = d_f D_{\min}^{d_f} D^{-(d_f+1)}$ is the probability density function of nanoparticles in nanofluids. Patterned after the probability theory, the probability density function $f(D)$ should satisfy the following normalization relationship or total cumulative probability (Yu and Cheng 2002a):

$$\int_{D_{\min}}^{D_{\max}} f(D)dD = 1 - \beta^{d_f} \equiv 1 \tag{2.7}$$

In general, most fractal objects have $\beta \le 10^{-2}$ (Cai et al. 2010; Cai and Sun 2013); thus, Equations 2.2 through 2.7 approximately hold. Thus, the fractal theory and technique can be used to analyze the characters of nanoparticles. The preceding equation holds if and only if the following equation is satisfied:

$$\beta^{d_f} \cong 0 \tag{2.8}$$

Equation 2.8 can be considered a criterion to decide whether nanoparticles can be characterized by fractal theory and technique.

Equations 2.2 through 2.8 form the basis for the analysis of nanofluids and will be employed to derive the heat transfer model in the following sections.

2.3 Modeling for Effective Thermal Conductivity of Nanofluids Using a Fractal Approach

2.3.1 Fractal Model

It is shown that the addition of nanoparticles enhances single-phase convection heat transfer (Xiao et al. 2009). Besides, the nanoparticles moving in fluids carry energy, and the heat exchange may occur between hot and cold regions transfer (Xiao et al. 2009, 2010). In this work, we assume that the enhancement of the thermal conductivity of nanofluids may be explained by the following two possible mechanisms. The first one is attributed to suspended nanoparticles in the base fluids, thus affecting the energy transport process, which corresponds to the conventional prediction. The second aspect is the heat convection caused by the Brownian motion of nanoparticles. Thus, the enhanced effective thermal conductivity of nanofluids (k_{eff}) can be expressed as

$$k_{eff} = k_c + k_s \tag{2.9}$$

where
  k_c is the thermal conductivity by heat convection caused by the Brownian motion of nanoparticles
  k_s is the thermal conductivity by stationary nanoparticles in the liquids, which is simulated by the Maxwell model (Maxwell 1873)

$$\frac{k_s}{k_f} = \frac{k_n - 2\phi(k_f - k_n) + 2k_f}{k_n + \phi(k_f - k_n) + 2k_f} \tag{2.10}$$

where k_f and k_n are the thermal conductivities of base fluid and particles, respectively.
  The focus of this work is to deduce the thermal conductivity by heat convection caused by the Brownian motion of nanoparticles. Since the nanoparticles are very small in nanofluids, the Brownian movement of the nanoparticles is probably caused by small-scale particles. The root-mean-square velocity (u) of a Brownian nanoparticle can be calculated as (Koo and Kleinstreuer 2004; Xiao et al. 2009)

$$u = \sqrt{\frac{3k_B T}{m}} = \frac{1}{D}\sqrt{\frac{18k_B T}{\pi \rho_n D}} \tag{2.11}$$

where
  $k_B = 1.38 \times 10^{-23}$ J/K is the Boltzmann constant
  T is the temperature
  m is the nanoparticle mass
  ρ_n is the density of nanoparticle
  D is the diameter of nanoparticle

The Reynolds number is defined by

$$Re = \frac{u \cdot D}{\upsilon} \tag{2.12}$$

where
  υ is the kinematic viscosity of base fluids
  u is the velocity of nanoparticle

Inserting Equation 2.11 into Equation 2.12, Equation 2.12 can be further reduced to

$$Re = \frac{1}{\upsilon}\sqrt{\frac{18k_B T}{\pi \rho_n D}} \tag{2.13}$$

For convection in the Stokes region, the average heat transfer coefficient h for an isothermal moving particle is given by (Acrivos and Taylor 1962; Prasher et al. 2005)

$$h = \frac{Nu \cdot k_f}{D} = \frac{(2 + 0.5RePr)k_f}{D} \tag{2.14}$$

where

Pr is the Prandtl number of fluids

Nu (=$2 + 0.5RePr$) (Acrivos and Taylor 1962) is the Nusselt number, which is often used to describe the heat transfer coefficient between nanoparticles and base fluids in nanofluids for fluids flowing around a single sphere (Acrivos and Taylor 1962; Jang and Choi 2004)

The Prandtl number of fluids is defined by

$$Pr = \frac{\upsilon}{\alpha} \tag{2.15}$$

where α is thermal diffusion coefficient.

Substituting Equations 2.13 and 2.15 into Equation 2.14 yields

$$h = \frac{(2 + 0.5RePr)k_f}{D} = \frac{k_f}{D}\left(2 + \frac{3}{2\alpha}\sqrt{\frac{2k_BT}{\pi\rho_n D}}\right) \tag{2.16}$$

Equation 2.16 depicts that the smaller the nanoparticles the larger the heat transfer coefficient by convection for nanoparticles moving in fluids. This phenomenon can be explained by the theory of Brownian motion: the smaller the size of nanoparticles in the fluid, the higher the velocity of nanoparticles' Brownian motion. Thus, the heat transferred by heat convection is enhanced.

The quantity of heat caused by a single nanoparticle moving in fluids can usually be expressed as

$$\psi = hA(T_p - T_f) \tag{2.17}$$

where

T_p and T_f are the temperatures of nanoparticles and fluids, respectively

A is the surface area of a nanoparticle whose diameter is D

The surface area of a nanoparticle can be expressed as

$$A = \pi D^2 \tag{2.18}$$

According to Equations 2.3 and 2.17, the total heat transfer by convection of all nanoparticles in nanofluids from D_{min} to D_{max} can be obtained as

$$Q_c = \int\limits_{D_{min}}^{D_{max}} hA(T_p - T_f)(-dN)$$

$$= \delta_T k_f d_f \left[\frac{3}{\alpha} \sqrt{\frac{2\pi k_B T}{\rho_n}} \frac{\left(D_{min}^{1/2-d_f} - D_{max}^{1/2-d_f}\right)}{2d_f - 1} + \frac{2\pi\left(D_{min}^{1-d_f} - D_{max}^{1-d_f}\right)}{d_f - 1} \right] D_{max}^{d_f} \frac{\Delta T}{\delta_T} \quad (2.19)$$

where $\delta_T \sim (\delta/Pr)$ is the thickness of the thermal boundary layer of heat convection caused by the Brownian motion (Jang and Choi 2004), and $\Delta T = T_p - T_f$. This work assumes that the temperature of nanofluids is moderate, not high. If the temperature is not high, the thermal conductivities k_c and k_s in Equation 2.9 may be independent of the temperature or the temperature difference. Equation 2.19 can be approximately applied to describe the heat exchange through convection between fluids and nanoparticles. Thus, the constant ratio of $\Delta T/\delta_T$ is assumed in the following derivation. It is postulated that the nanolayer of ordered liquid molecules is approximately considered as a hydrodynamic boundary layer (Yu et al. 1999; Jang and Choi 2004). Considering that the base fluid molecules at the interface have a much lower velocity than that of free molecules in the bulk fluids, we think the assumption is reasonable. So, the laminar boundary layer in Equation 2.19 can be approximated to $\delta = 3D_f$ (Yu et al. 1999; Jang and Choi 2004) due to the local thermal equilibrium assumed, where D_f is the diameter of liquid molecule.

With the aid of Equation 2.4, the total surface area of all nanoparticles can be obtained from Equation 2.18 as

$$A_t = \int\limits_{D_{min}}^{D_{max}} \pi D^2 \left[d_f D_{max}^{d_f} D^{-(d_f+1)} dD \right] = \frac{\pi\left(D_{max}^{2-d_f} - D_{min}^{2-d_f}\right) d_f D_{max}^{d_f}}{2 - d_f} \quad (2.20)$$

Since it is assumed that the temperature of nanofluids is not high and k_c in Equation 2.9 is independent of the temperature or the temperature difference, the equivalent thermal conductivity contributed by heat convection is approximated by

$$k_c = \frac{Q_c}{A_t(\Delta T/\delta_T)} = \frac{CD_f k_f \left[\frac{3}{\alpha} \sqrt{\frac{2k_B T}{\pi\rho_n}} \frac{(\beta^{1/2-d_f} - 1)}{(2d_f - 1)} + \frac{2(\beta^{1-d_f} - 1)D_{max}^{1/2}}{(d_f - 1)} \right]}{Pr(1 - \beta^{2-d_f})(2 - d_f)^{-1} D_{max}^{3/2}} \quad (2.21)$$

where C is an empirical constant, which is relevant to the thickness of thermal boundary layer, that is, $\delta_T \sim (\delta/Pr)$.

The average diameter of nanoparticles (D_{av}) can be obtained as (Xiao et al. 2013b)

$$D_{\max} \cong \left(\frac{4-d_f}{d_f} \right)^{1/4} D_{av} \qquad (2.22)$$

Equation 2.22 shows a fractal expression of the average diameter of nanoparticles.

By inserting Equation 2.22 into Equation 2.21, the equivalent thermal conductivity contributed by heat convection can be further modified as

$$k_c = \frac{CD_f k_f \left[\dfrac{3}{\alpha} \sqrt{\dfrac{2k_B T}{\pi \rho_n}} \dfrac{(\beta^{1/2-d_f}-1)d_f^{1/8}}{(2d_f-1)} + \dfrac{2(\beta^{1-d_f}-1)(4-d_f)^{1/8}D_{av}^{1/2}}{(d_f-1)} \right]}{Pr(1-\beta^{2-d_f})(4-d_f)^{3/8}(2-d_f)^{-1}d_f^{-1/4}D_{av}^{3/2}} \qquad (2.23)$$

The dimensionless effective thermal conductivity of nanofluids can be expressed as

$$k_{eff}^+ = \frac{k_{eff}}{k_f} = \frac{k_s + k_c}{k_f} = k_s^+ + k_c^+ \qquad (2.24a)$$

where

$$k_s^+ = \frac{k_s}{k_f} = \frac{k_n - 2\phi(k_f - k_n) + 2k_f}{k_n + \phi(k_f - k_n) + 2k_f} \qquad (2.24b)$$

$$k_c^+ = \frac{k_c}{k_f} = \frac{CD_f \left[\dfrac{3}{\alpha} \sqrt{\dfrac{2k_B T}{\pi \rho_n}} \dfrac{(\beta^{1/2-d_f}-1)d_f^{1/8}}{(2d_f-1)} + \dfrac{2(\beta^{1-d_f}-1)(4-d_f)^{1/8}D_{av}^{1/2}}{(d_f-1)} \right]}{Pr(1-\beta^{2-d_f})(4-d_f)^{3/8}(2-d_f)^{-1}d_f^{-1/4}D_{av}^{3/2}} \qquad (2.24c)$$

Equation 2.24 is the fractal analytical expressions for the effective thermal conductivity of nanofluids. Equation 2.24 indicates that the effective thermal conductivity of nanofluids is expressed as a function of the thermal conductivities of the base fluid and nanoparticles, the average diameter of nanoparticles, the nanoparticles concentration, the fractal dimension of nanoparticles, and the physical properties of fluids. Equation 2.24c denotes that the effective thermal conductivity decreases with the increase of the average size of nanoparticles. This also means that if the average size of nanoparticles is small, it causes the increase of nanoparticles moving in fluids, leading to the increase in the heat transfer of nanofluids. It is expected that Equation 2.24 has less empirical constants and every parameter in Equation 2.24 has clear physical meaning. The proposed fractal model can reveal the physical mechanisms of the heat transfer of nanofluids.

2.3.2 Results and Discussions

Figure 2.4 shows that the fractal dimension (by Equation 2.2) of nanoparticles increases as the volumetric nanoparticle concentration (ϕ) increases. This means that the high ϕ causes the increase of the number of nanoparticles, leading to the increase of the fractal dimension of nanoparticles.

Figure 2.5 compares the effective thermal conductivity of nanofluids from the present model (Equation 2.24) and that from reported experiments (Masuda et al. 1993; Xie et al. 2002; Das et al. 2003b) for different nanoparticle–water suspensions. In this calculation, empirical parameter $C = 236$ is selected (Xiao et al. 2013b). It is seen that the predicted thermal conductivities by fractal technique are in good agreement with experimental data. It can also be found that the effective thermal conductivity of nanofluids increases with the increase in nanoparticle concentration. As shown in Figure 2.5, the effective thermal conductivity of nanofluids for nanoparticles smaller in size is larger than that of the bigger-sized nanoparticles at the given concentration. This means the smaller size of the nanoparticles at lower concentration causes increase in nanoparticles moving in fluids, which leads to the increase of the effective thermal conductivity from convection.

Figure 2.6 depicts the average diameter of nanoparticles dependence of the effective thermal conductivity of nanofluids. It can be seen from Figure 2.6 that the effective thermal conductivity ($k_{eff}^+ = k_{eff}/k_f$) decreases with the increase of the average diameter of nanoparticles for CuO–water nanofluids. It is also interestingly seen from Figure 2.6 that when the average diameter of nanoparticles is less than about 16 nm, the effective thermal conductivity increases drastically. This reveals that the convection due to the Brownian

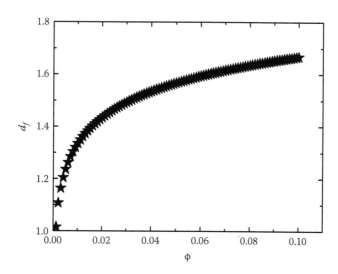

FIGURE 2.4
The fractal dimension versus the volumetric concentration of nanoparticles.

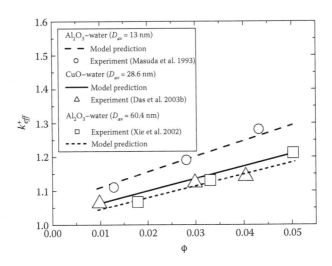

FIGURE 2.5
Comparison of the proposed model and current experimental data (Masuda et al. 1993; Xie et al. 2002; Das et al. 2003b) for different nanofluids.

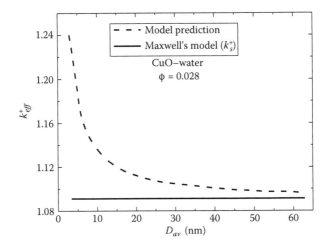

FIGURE 2.6
Dependence of the thermal conductivity enhancement on nanoparticle diameter for the CuO–water suspension at $\phi = 0.028$.

motion of nanoparticles has significant influence on the thermal conductivity of nanofluids when the average diameter of nanoparticles is less than about 16 nm. This phenomenon can be interpreted to mean that the small size of nanoparticles causes the increase in the Brownian motion of the nanoparticles in the fluids. However, the Maxwell model is not relevant to the average size of nanoparticles.

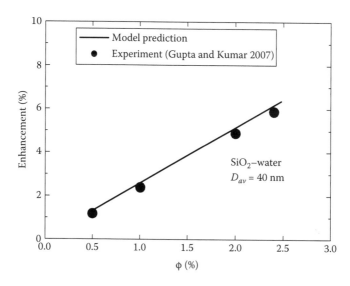

FIGURE 2.7
Increase in thermal conductivity for silica nanoparticle in water.

Figure 2.7 shows a comparison of increase in thermal conductivity due to the Brownian motion by the present model and that from reported experiments (Gupta and Kumar 2007) for silica nanoparticles in water. In the figure, an increase in the net thermal conductivity compared to the pure fluids, which increases with nanoparticle concentration, is shown. Besides, it can be seen that the conductivity due to the Brownian motion is not negligible but not as high as the predicted thermal conductivity. These trends are consistent with the experimental results (Gupta and Kumar 2007; Shima et al. 2009).

2.3.3 Summary and Conclusions

In this work, taking into account the heat convection between nanoparticles and liquids due to the Brownian motion of nanoparticles in fluids, the analytical expression for effective thermal conductivity of nanofluids is developed based on the fractal geometry theory. The proposed model is expressed as a function of the thermal conductivities of the base fluids and the nanoparticles, the average diameter of nanoparticles, the nanoparticle concentration, the fractal dimension of nanoparticles, and physical properties of fluids. Based on the parametric effect analysis, we conclude that the effective thermal conductivity of nanofluids is negatively correlated with the average size of nanoparticles but positively correlated with the nanoparticle concentration. Every parameter of the proposed formulas on the effective thermal conductivity of nanofluids has clear physical meaning. The model predictions are compared with the existing experimental data and excellent agreement between the model predictions and experimental data is found. The validity of the present model is thus verified. The proposed fractal model can reveal the physical mechanisms of heat transfer of nanofluids.

2.4 Modeling for Convective Heat Transfer of Nanofluids Based on Fractal Theory and Monte Carlo Simulation

2.4.1 Formulation of the Proposed Model

The heat exchange of nanofluids is mainly through convection at natural convection region. And the nanoparticles moving in fluids carry energy, and the heat exchange may occur between hot and cold regions. Thus, the heat transferred by convection for a single nanoparticle moving in fluids can be expressed as (Xiao et al. 2013a)

$$Q_{D_i} = hA_{D_i}\Delta T \tag{2.25}$$

where
ΔT is the wall superheat
h is the mean heat transfer coefficient, which is given by Equation 2.16

Substituting Equations 2.16 and 2.18 into Equation 2.25 yields

$$Q_{D_i} = \frac{k_f(2+0.5RePr)}{D_i}\pi D_i^2\Delta T = \pi k_f D_i\left(2+\frac{3}{2\alpha}\sqrt{\frac{2k_BT}{\pi\rho_nD_i}}\right)\Delta T \tag{2.26}$$

Experimental observations (Hong et al. 2006) have shown that the nanoparticles may have different sizes in nanofluids. Thus, the total heat transfer by convection of all nanoparticles in nanofluids can be obtained by adding the heat transfer by individual particle as

$$Q_{t,c} = \sum_{i=1}^{J}Q_{D_i} = \sum_{i=1}^{J}\pi kD_i\left(2+\frac{3}{2\alpha}\sqrt{\frac{2k_BT}{\pi\rho_nD_i}}\right)\Delta T \tag{2.27}$$

The total surface area of all nanoparticles can be obtained with the aid of Equation 2.18:

$$A_t = \sum_{i=1}^{J}\pi D_i^2 \tag{2.28}$$

According to Equations 2.27 and 2.28, the total heat flux caused by all nanoparticles moving in fluids can be expressed as

$$q_{t,c} = \frac{Q_{t,c}}{A_t} = \frac{k_f \sum_{i=1}^{J} D_i \left(2 + \frac{3}{2\alpha} \sqrt{\frac{2k_B T}{\pi \rho_n D_i}}\right) \Delta T}{\sum_{i=1}^{J} D_i^2} \tag{2.29}$$

We divide total convective heat transfer of nanofluids into two portions: the heat transferred ($q_{t,c}$) by heat convection caused by the Brownian motion of nanoparticles and the heat transferred ($q_{b,c}$) by natural convection from the base fluids. Thus, the total heat flux of nanofluids by convection can be expressed as (Xiao et al. 2013a)

$$q_{t,nc} = q_{t,c} + q_{b,c} \tag{2.30}$$

Past investigations (Xiao and Yu 2007a,b) have found that there is a significant effect of surface characteristics on boiling performance and mechanisms, when a pure liquid was boiled over heating surfaces. Actually, the transport properties of the heater affect the extent of the thermal interaction among the cavities, causing activation and deactivation of individual cavities.

The heat flux by natural convection from the base fluids in saturation fluids is usually defined as

$$q_{b,c} = h_{b,c} \Delta T \tag{2.31}$$

where $h_{b,c}$ is the average heat transfer coefficient by natural convection from the base fluids, which is given by (Han and Griffith 1965)

$$h_{b,c} = 0.14 \rho_f c_p \left[\frac{\gamma_1 g (T_w - T_l) \alpha^2}{\upsilon} \right]^{1/3} \tag{2.32a}$$

for turbulent range where $2 \times 10^7 < Ra < 3 \times 10^{10}$. For laminar range where $10^5 < Ra < 2 \times 10^7$, $h_{b,c}$ is given by

$$h_{b,c} = 0.54 \rho_f c_p \left[\frac{\gamma_1 g (T_w - T_l) \alpha^3}{\sqrt{A_h} \upsilon} \right]^{1/4} \tag{2.32b}$$

where
  A_h is the area of heating surface
  ρ_f is the base fluids density
  c_p is the specific heat at constant pressure
  γ_1 is the volumetric thermal expansion coefficient of liquid
  g is the acceleration due to gravity
  T_w is wall temperature
  T_l is bulk temperature

Inserting Equations 2.29 and 2.31 into Equation 2.30, the total heat flux from convective heat transfer of nanofluids is then

$$q_{t,nc} = q_{t,c} + q_{b,c} = \left[\frac{k_f \sum_{i=1}^{J} D_i \left(2 + \frac{3}{2\alpha} \sqrt{\frac{2k_B T}{\pi \rho_n D_i}} \right)}{\sum_{i=1}^{J} D_i^2} + h_{b,c} \right] \Delta T \qquad (2.33)$$

Equation 2.33 relates the total heat flux from convective heat transfer of nanofluids to the parameters of nanofluid, such as the nanoparticle sizes (D_i), the thermal conductivities of base fluids (k_f), and the wall superheat (ΔT), as well as fluid properties ($\alpha, \rho, h_{b,c}$). In Equation 2.33, ΔT is a variable. Equation 2.33 takes into account the effect of convection caused by the Brownian motion. In this model, no new empirical constant is introduced.

2.4.2 Methodology for the Monte Carlo Technique

Besides Equation 2.22, the mean nanoparticle diameter D_{av} can also be calculated as

$$D_{av} = \int_{D_{min}}^{D_{max}} D_i f(D_i) dD_i = \int_{D_{min}}^{D_{max}} D_i \left[d_f D_{min}^{d_f} D_i^{-(d_f+1)} \right] dD_i \cong \frac{d_f}{d_f - 1} D_{min} \qquad (2.34)$$

So, given the mean diameter, the minimum and maximum diameters of nanoparticles can be obtained by

$$D_{min} = \frac{d_f - 1}{d_f} D_{av} \qquad (2.35a)$$

$$D_{max} = D_{av} \frac{d_f - 1}{d_f} \left(\frac{D_{max}}{D_{min}} \right) = \frac{d_f - 1}{\beta d_f} D_{av} \qquad (2.35b)$$

The cumulative probability in the range of $D_{min} \sim D_i$ can be obtained from the probability density function $f(D_i) = d_f D_{min}^{d_f} D_i^{-(d_f+1)}$ as

$$R_i(D_i) = \int_{D_{min}}^{D_i} f(D_i) dD_i = \int_{D_{min}}^{D_i} d_f D_{min}^{d_f} D_i^{-(d_f+1)} dD_i = 1 - \left(\frac{D_{min}}{D_i} \right)^{d_f} \qquad (2.36)$$

Equation 2.36 indicates that $R=0$ as $D \rightarrow D_{min}$ and $R \approx 1$ as $D \rightarrow D_{max}$. R in Equation 2.36 is in the range of 0–1.

For the ith nanoparticles chosen randomly, from Equation 2.36, the diameter D_i is expressed as

$$D_i = \frac{D_{min}}{(1-R_i)^{1/d_f}} = \beta \frac{D_{max}}{(1-R_i)^{1/d_f}} \tag{2.37}$$

where $i = 1, 2, 3, \ldots, J$, and J is the total number of Monte Carlo simulations in one run for a given concentration. Equation 2.37 presents an explicit probability model for nanoparticles size distribution in the present simulation. Equation 2.37 denotes that since R_i is a random number of 0–1 produced by computer, the nanoparticle diameter D_i is determined randomly, and this also simulates the randomness and fractal distribution of nanoparticle sizes.

The mean diameter of all nanoparticles calculated in the present Monte Carlo simulations can be written as

$$\bar{D} = \frac{1}{J} \sum_{i=1}^{J} D_i \tag{2.38}$$

The algorithm for the determination of the convective heat transfer of nanofluids is summarized as follows:

1. Given are a volumetric nanoparticle concentration ϕ and a mean nanoparticle diameter D_{av} in simulations.
2. Find d_f, D_{min}, and D_{max} from Equations 2.2, 2.35a, and 2.35b, respectively.
3. Produce a random number R_i of 0–1 using computer.
4. Calculate D_i from Equation 2.37.
5. If $D_i > D_{max}$ or $D_i < D_{min}$, return to procedure 3; otherwise, continue to the next procedure.
6. Find $\bar{D}$ from Equation 2.38.
7. Find the total heat flux ($q_{t,c}$) caused by all nanoparticles moving in fluids from Equation 2.29 and the heat flux ($q_{b,c}$) from the base fluids by natural convection from Equation 2.31.
8. Find the total heat flux from the convective heat transfer of nanofluids $q_{t,nc}$ from Equation 2.33.

Procedures 4–8 are repeated for the calculation of total heat flux from the convective heat transfer of nanofluids until a converged value is obtained at a given concentration. The convergence criterion is that when the following condition is satisfied, that is,

$$\bar{D} = \frac{1}{J} \sum_{i=1}^{J} D_i \geq D_{av} \tag{2.39}$$

stop the simulation and record the final $q_{t,nc}$ and the total number (J) in one run for a given concentration. In Equation 2.39, D_{av} is calculated from Equation 2.34. If the converged heat flux from the convective heat transfer of nanofluids is obtained in one run, set the heat flux from the convective heat transfer of nanofluids as $q_{t,nc}^{(n)}$ ($n = 1, 2, 3, ..., M$). Then, the mean $q_{t,nc}$ for a given volumetric nanoparticle concentration (ϕ) is calculated from

$$\langle q_{t,nc} \rangle = \frac{1}{M} \sum_{n=1}^{M} q_{t,nc}^{(n)} \tag{2.40}$$

where M is the total number of runs for a given concentration.

The variance is defined by

$$\sigma = \sqrt{\langle q_{t,nc}^2 \rangle - \langle q_{t,nc} \rangle^2} \tag{2.41}$$

where

$$\langle q_{t,nc}^2 \rangle = \frac{1}{M} \sum_{n=1}^{M} q_{t,nc}^{2(n)} \tag{2.42}$$

From this algorithm for predicting the total heat flux from the convective heat transfer of nanofluids by the Monte Carlo method, it is seen that the present algorithm for the total heat flux from the convective heat transfer of nanofluids is quite simple. This technique reveals the randomness and the fractal distribution of nanoparticle sizes.

2.4.3 Results and Discussions

Boiling curves (Das et al. 2003a; Bang and Chang 2005) of Al_2O_3–water nano-fluids at the low heat flux of natural convection is studied at $D_{av} = 47$ nm and $\phi = 1\%$. Figure 2.8 compares the heat flux from convective heat transfer of Al_2O_3–water nanofluids by the Monte Carlo simulations with the experimental data (Das et al. 2003a; Bang and Chang 2005). It is seen from Figure 2.8 that the predicted heat flux by the present Monte Carlo technique is in good agreement with the experimental data (Das et al. 2003a; Bang and Chang 2005) at $D_{av} = 47$ nm and $\phi = 1\%$. Das et al. (2003a) conducted experimental studies of pool boiling characteristics of Al_2O_3–water nanofluid at $D_{av} = 38$ nm and $\phi = 0.1\%$ under atmospheric conditions. The heat flux (Das et al. 2003a) ranged from 2×10^4 to 1.2×10^5 W/m² at the natural convection stage. Figure 2.9 is a comparison between the simulated heat flux from convective heat transfer and the experimental data (Das et al. 2003a) at $D_{av} = 38$ nm and $\phi = 0.1\%$. Figure 2.9 indicates that the predicted heat flux of nanofluids is well in accord with the experimental measurements (Das et al. 2003a). The deviation between the simulated results and the experiment data is very small. A good

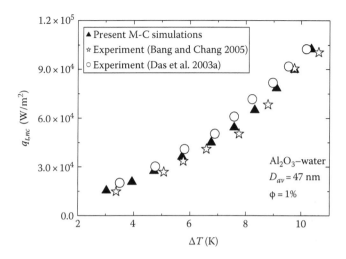

FIGURE 2.8
A comparison between the present Monte Carlo simulations and the experimental data at $D_{av} = 47$ nm and $\phi = 1\%$ for Al_2O_3 nanofluids.

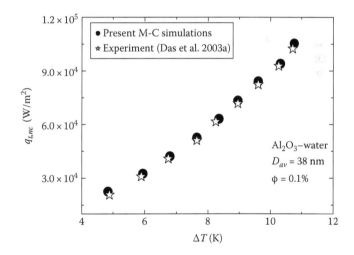

FIGURE 2.9
A comparison between the present Monte Carlo simulations and the experimental data at $D_{av} = 38$ nm and $\phi = 0.1\%$ for Al_2O_3 nanofluids.

agreement between the predicted heat flux by the present Monte Carlo technique and the experimental data is again found. It can be seen from Figures 2.8 and 2.9 that the heat flux from convective heat transfer increases with ΔT. This means that a higher temperature may cause a stronger Brownian motion, and, thus, may contribute more to the heat transfer from convection.

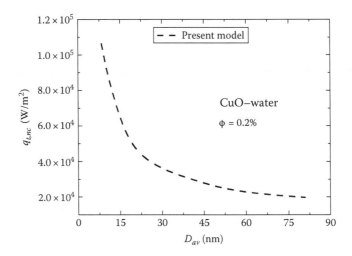

FIGURE 2.10
The heat flux from convective heat transfer of CuO nanofluids versus the average diameter of nanoparticles.

Figure 2.10 shows the heat flux from the convective heat transfer of CuO nanofluids versus the average diameter of nanoparticles at $\phi=0.2\%$. It can be seen from Figure 2.10 that the heat flux at natural convection stage decreases with an increase in the average size of nanoparticles. This phenomenon can be explained by the theory of Brownian motion: the smaller the average size of nanoparticles in the fluids, the higher the velocity of nanoparticles' Brownian motion. Thus, the heat transferred by heat convection is enhanced. These are all expected and consistent with the practical physical phenomena. Therefore, the Monte Carlo simulations can reveal the detailed mechanisms of heat transfer at convection region in nanofluids.

2.4.4 Summary and Conclusions

The Monte Carlo technique combined with fractal geometry theory is applied to predict the convective heat transfer of nanofluids in this chapter. This work also takes into account the convection caused by the Brownian motion and the fractal distribution of nanoparticle sizes. The proposed model is explicitly related to the nanoparticle sizes, the volumetric nanoparticle concentration, the thermal conductivity of base fluids, the fractal dimension of nanoparticles, the size of nanoparticles, and the temperature, as well as random number. Based on the parametric effect analysis, we conclude that the convective heat transfer of nanofluids is negatively correlated with the average size of nanoparticles but positively correlated with the wall superheat. This model has the characters of both analytical and numerical solutions. The model predictions are compared with the existing experimental data, and good agreement between the model predictions and experimental data is found. The validity of the present model is thus verified. All parameters of the proposed formulas on the convective heat transfer of nanofluids have clear physical meaning. The proposed model can reveal the physical mechanisms of the convection heat transfer of nanofluids. Besides the analytical and numerical methods, the present technique might provide us with an approach that might also have the potential in the analysis of the transport properties such as the magnetic and electrical properties of nanofluids.

2.5 Fractal Model for Critical Heat Flux of Nanofluids

2.5.1 Fractal Method and Calculations

The critical heat flux (CHF) of nucleate pool boiling heat transfer in nanofluids is shown in Figure 2.11 (Kim et al. 2007b).

In base fluids, it is generally recognized that the main mechanism contributing to nucleate boiling heat transfer is the bubble generation and departure from the active cavity on the superheated surface in CHF region. Thus, there are two main mechanisms contributing to nucleate pool boiling heat transfer of nanofluids in the CHF region: one is the heat ($q_{t,c}$) transferred by the heat convection caused by the Brownian motion of nanoparticles and the other is the heat ($q_{b,CHF}$) transferred by the bubbles generation and departure from the base fluids in the CHF region (see Figure 2.11). In the CHF region, the total heat flux of nucleate pool boiling heat transfer of nanofluids can be expressed as (Xiao 2013)

$$q_{N,CHF} = q_{t,c} + q_{b,CHF} \tag{2.43}$$

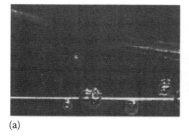

(a)   (b)

FIGURE 2.11

SEM micrographs (Kim et al. 2007b) of CHF region in nanofluids: (a) CHF$=0.5 \times 10^6$ W/m^2 and (b) CHF$=1.0 \times 10^6$ W/m^2.

The heat flux from all nanoparticles moving in liquids can be expressed as (Xiao 2013)

$$q_{t,c} = \frac{d_f^{1/4}(4-2d_f)k_f[k_n(1+2\phi)+2k_f(1-\phi)](\beta^{-d_N+1}-1)\Delta T}{(d_f-1)(4-d_f)^{1/4}[k_n(1-\phi)+k_f(2+\phi)](1-\beta^{-d_N+2})D_{av}}$$

(2.44)

Equation 2.44 is the fractal analytical expression for the heat transferred by nanoparticles moving in nanofluids. Equation 2.44 shows that the heat flux ($q_{t,c}$) increases as the average diameter of nanoparticles D_{av} decreases. This is expected because the small diameter of nanoparticles may cause the increase in its velocity, leading to more heat transferred by nanoparticles moving in nanofluids. Equation 2.44 depicts that the heat flux ($q_{t,c}$) is a function of the average diameter of nanoparticles D_{av}, the volumetric nanoparticle concentration ϕ, the thermal conductivity of nanoparticles k_n, the fractal dimension of nanoparticles d_f, and the thermal conductivity of liquid k_f.

Several investigators have found that the surface characteristics affect the pool boiling performance and mechanisms when a pure liquid is boiled over heating surfaces (Yu and Cheng 2002b; Xiao and Yu 2007a,b). The density of active sites on the heater surface is affected by the interaction among several parameters on the heater and the liquid sides, as well as the liquid–solid contact angle (Yu and Cheng 2002b; Xiao and Yu 2007a,b). For example, the distribution of available cavities on the heater surface and the liquid–solid contact angle determine which cavities could potentially be active. At the same time, the transport properties of the heater affect the extent of the thermal interaction among the cavities, causing the activation and deactivation of individual cavities.

Xiao and Yu (2007a) have obtained the fractal analytical expressions of nucleate boiling heat transfer $q_{b,\mathrm{CHF}}$ from the base fluid in the CHF region. The CHF of pool boiling can be expressed as

$$q_{b,\mathrm{CHF}} = c_q \frac{d_{fc}}{d_{fc}+2} \frac{4\pi\alpha}{3} \left(\frac{\Delta T}{T_w-T_l}\right)^2 D_{c,\mathrm{max}}^{-2} \left[\left(\frac{D_{c,\mathrm{max}}}{D_{c,\mathrm{min}}}\right)^{d_{fc}+2}-1\right]$$

(2.45)

where
$c_q = (\pi/6)h_{fg}\rho_g D_b^3$ is the heat flux removed by a single bubble
h_{fg} is the latent heat of vaporization
ρ_g is the vapor density
$\Delta T = T_w - T_s$, T_s is the saturation temperature of liquids
D_b is the bubble departure diameter
$D_{c,\mathrm{max}}$ and $D_{c,\mathrm{min}}$ are, respectively, the maximum and minimum diameters of active cavity
d_{fc} is the fractal dimension of active cavity on the heated surface

The bubble departure diameter D_b can be obtained as (Mikic and Rohsenow 1969)

$$D_b = c_0 \left[\frac{\sigma}{g(\rho_f - \rho_g)} \right]^{1/2} Ja^{*5/4} \qquad (2.46)$$

where

$c_0 = 1.5 \times 10^{-4}$ for water and $c_0 = 4.65 \times 10^{-4}$ for the other liquid
σ is the surface tension of liquid
ρ_f is the density of liquid
g is the acceleration due to gravity
Ja^* is the Jakob number, which is given by

$$Ja^* = \frac{\rho_l c_{pl} T_s}{\rho_g h_{fg}} \qquad (2.47)$$

It is shown by Equations 2.45 and 2.46 that the boiling heat transfer from the base fluids in the CHF region increases with the increasing bubble departure diameter.

Since $D_{c,max}/D_{c,min} \geq 10^2$, $0 < d_{fc} < 2$, $0 < 2 + d_{fc} < 4$, and $(D_{c,max}/D_{c,min})^{d_{fc}+2} \ll 1$, Equation 2.45 can be further reduced to

$$q_{b,CHF} = c_q \frac{d_{fc}}{d_{fc}+2} \frac{4\pi\alpha}{3} \left(\frac{\Delta T}{T_w - T_l} \right)^2 D_{c,max}^{-2} \left(\frac{D_{c,max}}{D_{c,min}} \right)^{d_{fc}+2} \qquad (2.48)$$

The minimum active cavity diameter $D_{c,min}$ and the maximum active cavity diameter $D_{c,max}$ can be predicted by the model (Hsu 1962) as

$$D_{c,min} = \frac{2\delta}{C_1} \left[1 - \frac{\theta_s}{\theta_w} - \sqrt{\left(1 - \frac{\theta_s}{\theta_w} \right)^2 - \frac{4\zeta C_2}{\delta \theta_w}} \right] \qquad (2.49a)$$

$$D_{c,max} = \frac{2\delta}{C_1} \left[1 - \frac{\theta_s}{\theta_w} + \sqrt{\left(1 - \frac{\theta_s}{\theta_w} \right)^2 - \frac{4\zeta C_2}{\delta \theta_w}} \right] \qquad (2.49b)$$

where

$\zeta = 2\sigma T_s/(\rho_g h_{fg})$
$C_1 = (1 + \cos\theta)/\sin\theta$
$C_2 = 1 + \cos\theta$, with θ being the contact angle of the fluid and the heater material
$\theta_s = T_s - T_l$
$\theta_w = T_w - T_l$
δ is the thermal boundary layer thickness in nanofluid, which can be expressed as

$$\delta = \frac{k_{eff}}{h} \tag{2.50}$$

with h being the heat transfer coefficient obtained from Equation 2.16.

In nucleate pool boiling of the base fluids, the fractal dimension d_{fc} of active cavity on the heated surface is given by Yu and Cheng (2002b) as

$$d_{fc} = \frac{\ln[\bar{D}_{c,\max} / (\sqrt{2}D_{c,\min})]^2}{\ln(\gamma^{-1})} \tag{2.51}$$

where $\gamma = D_{c,\min}/D_{c,\max}$. Here $\bar{D}_{c,\max}$ is the value averaged over all the maximum active cavities

$$\bar{D}_{c,\max} = \frac{1}{(T_w - T_s)} \int_{T_s}^{T_w} D_{c,\max}(T_w)dT_w = \frac{1}{\Delta T}\sum_{j=1}^{m}D_{c,\max}(T_{w_j})\delta T_w = \frac{1}{m}\sum_{j=1}^{m}D_{c,\max}(T_{w_j})$$

$$\tag{2.52}$$

where $m = \Delta T / \delta T_w$, and δT_w is assumed to be a constant. In this equation, $T_{w_j} = T_s + j(\delta T_w)$ with $j = 1, 2, \ldots, m$. For example, if we choose $\delta T_w = 0.2°C$, then $m = 5$ for $\Delta T = 1°C$, and $m = 50$ for $\Delta T = 10°C$.

Inserting Equations 2.44 and 2.48 into Equation 2.43, we can obtain a fractal model for heat transfer of nanofluids in the CHF region as

$$q_{N,\,CHF} = q_{t,c} + q_{b,\,CHF} = \frac{d_f^{1/4}(4 - 2d_f)k_f[k_n(1+2\phi) + 2k_f(1-\phi)](\beta^{-d_N+1} - 1)\Delta T}{(d_f - 1)(4 - d_f)^{1/4}[k_n(1-\phi) + k_f(2+\phi)](1 - \beta^{-d_N+2})D_{av}}$$

$$+ c_q \frac{d_{fc}}{d_{fc}+2} \frac{4\pi\alpha}{3} D_{c,\max}^{-2}\left(\frac{D_{c,\max}}{D_{c,\min}}\right)^{d_{fc}+2}\left(\frac{\Delta T}{T_w - T_l}\right)^2 \tag{2.53}$$

where d_f and d_{fc} are obtained from Equations 2.2 and 2.51, respectively. Equation 2.53 is the fractal analytical expressions of CHF for pool boiling heat transfer in nanofluids and indicates that the CHF of pool boiling heat transfer in nanofluids is explicitly related to the average diameter of nanoparticles, the volumetric nanoparticle concentration, the thermal conductivity of nanoparticles, the fractal dimension of nanoparticles, the fractal dimension of active cavity on the heated surface, the temperature, and the properties of fluids. It is expected that Equation 2.53 has less empirical constants and each parameter in Equation 2.53 has a clear physical meaning. The proposed fractal model can reveal the mechanism of pool boiling heat transfer on CHF in nanofluids.

2.5.2 Results and Discussions

Figure 2.12 is a plot of the fractal dimension (by Equation 2.51) of active cavities on the heated surfaces versus wall superheat. According to the fractal geometry theory, the fractal dimension of active cavities should be in the range of $1 < d_{fc} < 2$ in two dimensions. Figure 2.12 shows that the fractal dimension of active cavities increases with wall superheat. This means that d_{fc} is a function of wall superheat. These are all expected and consistent with the practical physical phenomena.

In the CHF region, the pool boiling characteristics of water-based nanofluids with titania ($D_{av} = 45$ nm, $\phi = 0.1\%$) and alumina ($D_{av} = 47$ nm, $\phi = 0.1\%$) nanoparticles are investigated on a thermally heated disk heater at saturated temperature and atmospheric pressure (Kim et al. 2010). The water-based nanofluid is prepared by dispersing the dry powders into deionized, distilled water with 3 h ultrasonic vibration. Figure 2.13 shows a comparison between the CHF of pool boiling heat transfer from the present model (Equation 2.53) and that from reported experiments (Kim et al. 2010) at $D_{av} = 45$ nm and $\phi = 0.1\%$ for TiO_2 nanofluids. It is seen that there is a good agreement between the model prediction and the experimental data. The solid line in Figure 2.13 represents the predictions by the present model.

Figure 2.14 shows a comparison between the CHF predicted by the present model and that from reported experiments (Kim et al. 2010) at $D_{av} = 47$ nm and $\phi = 0.1\%$ for Al_2O_3 nanofluids. It is seen that an excellent agreement between the model predictions and the experimental data is again found.

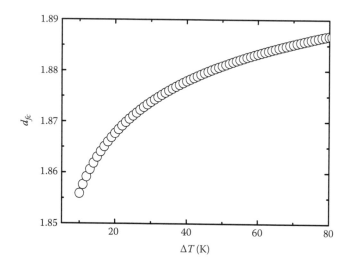

FIGURE 2.12
The fractal dimension of active cavities on the heated surfaces versus wall superheat.

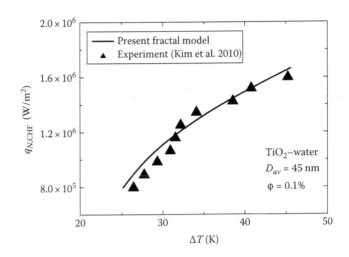

FIGURE 2.13
Comparison between the present model predictions and the experimental data at $D_{av} = 45\,nm$ and $\phi = 0.1\%$ for TiO$_2$ nanofluids.

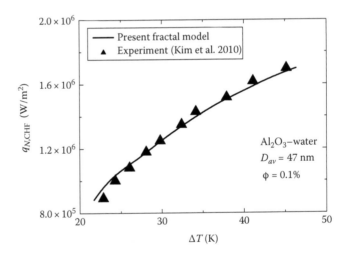

FIGURE 2.14
Comparison between the present model predictions and the experimental data at $D_{av} = 47\,nm$ and $\phi = 0.1\%$ for Al$_2$O$_3$ nanofluids.

Liu and Liao (2008) conducted experiments on CHF in pool boiling under atmospheric pressure at $D_{av} = 35\,nm$ and $\phi = 0.5\%$ for SiO$_2$ nanofluids. The pool nucleate boiling heat transfer experiments of water-based nanoparticle suspensions on the plain heated copper surface are carried out. Figure 2.15 shows a comparison between the model predictions and

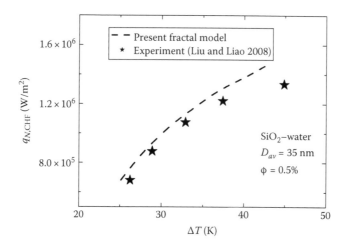

FIGURE 2.15
Comparison between the model prediction and experimental data at $D_{av}=35$ nm and $\phi=0.5\%$ for SiO$_2$ nanofluids.

experimental data (Liu and Liao 2008) at $D_{av}=35$ nm and $\phi=0.5\%$ for SiO$_2$ nanofluids. Again, a reasonable agreement is observed between experimental and predicted values.

To investigate the CHF characteristics of nanofluids, the pool boiling experiments of nanofluids with various concentrations of Al$_2$O$_3$ nanoparticles ($D_{av}=47$ nm) are carried out using a 0.2 mm diameter cylindrical Ni–Cr wire under atmospheric pressure (Kim et al. 2007a). In Figure 2.16,

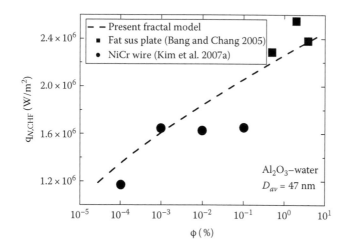

FIGURE 2.16
Comparison of predicted fractal models and existing experimental data at $D_{av}=47$ nm for Al$_2$O$_3$ nanofluids.

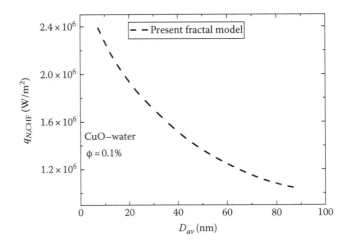

FIGURE 2.17
CHF of pool boiling heat transfer versus the average diameter of nanoparticles at $\phi=0.1\%$ in CuO nanofluids.

the results from Equation 2.53 are compared with the present prediction results and the data collected by others (Bang and Chang 2005; Kim et al. 2007a) over a wide range of volumetric nanoparticle concentrations at $D_{av}=47$ nm for Al_2O_3 nanofluids. As can be seen from Figure 2.16, the results from the model are in fair agreement with experimental data in the literature. The data presented by Bang and Chang (2005) are higher than the others because the CHF of pure water in their experiment is higher than that in the other studies shown in Figure 2.16. As a result, the maximum CHF increments in all work on nanofluids are about 900 kW/m². This shows that the result of our study is consistent with those of other work under atmospheric pressure.

Figure 2.17 shows the CHF of pool boiling heat transfer versus the average diameter of nanoparticles at $\phi=0.1\%$ in CuO nanofluids. It can be seen from Figure 2.17 that the CHF decreases with the increase in the average diameter of nanoparticles. This can be explained by the fact that the small nanoparticles cause the increase in nanoparticles moving in liquid, leading to the increase of heat transfer of nanofluids. These are all expected and consistent with the practical physical phenomena. Therefore, the present fractal model can reveal the mechanism of heat transfer of nanofluids in the CHF region.

2.5.3 Summary and Conclusions

With the consideration of nanoparticles moving in liquids, we derive the analytical expressions for pool boiling heat transfer of nanofluids in the CHF region based on the fractal geometry theory in this work. In the present approach, the CHF of nanofluids is found to be a function of the average diameter of nanoparticles, the volumetric nanoparticle concentration, the thermal conductivity of nanoparticles, the fractal dimension of nanoparticles, the fractal dimension of active cavity on the heated surface, the temperature, and the properties of fluids. From the parametric effect analysis, we conclude that the CHF of nanofluids is negatively correlated with the average diameter of nanoparticles but positively correlated with the volumetric nanoparticle concentration. So the proposed fractal model can reveal the mechanism of pool boiling heat transfer on CHF in nanofluids. The analytical solution CHF model of nanofluids does not have any empirical constant, and each parameter in this model has a specific physical meaning. The model predictions are compared with the existing experimental data, and a good agreement is found. The validity of the present fractal model is thus verified.

2.6 Fractal Analysis of Pool Boiling Heat Transfer in Nanofluids

2.6.1 Modeling Heat Flux of Subcooled Pool Boiling in Fractal Nanofluids

In general, there are two main mechanisms contributing to subcooled pool boiling heat transfer of nanofluids: the heat flux ($q_{t,c}$) from all nanoparticles moving in liquid and the other (q_{sb}) from subcooled pool boiling of the base fluids. Thus, the heat flux (q_{sn}) of subcooled pool boiling heat transfer of nanofluids can be expressed as (Xiao et al. 2014)

$$q_{sn} = q_{t,c} + q_{sb} \tag{2.54}$$

According to Equations 2.16 and 2.18, the heat flux from all nanoparticles moving in liquids can be further modified as (Xiao et al. 2014)

$$q_{t,c} = \frac{\Delta T \int_{D_{min}}^{D_{max}} hA(-dN)}{\int_{D_{min}}^{D_{max}} A(-dN)}$$

$$= \left\{ \frac{3}{\alpha} \sqrt{\frac{2k_B T}{\pi \rho_n}} \frac{[\beta^{(\log_\beta^\phi - 1.5)} - 1]}{(3 - 2\log_\beta^\phi)} \left[\frac{\beta(2 - \log_\beta^\phi)(1 - \beta^{(1 - \log_\beta^\phi)})}{(1 - \log_\beta^\phi)} \right]^{3/2} D_{av}^{-1/2} \right.$$

$$\left. + \frac{2\beta(2 - \log_\beta^\phi)[\beta^{(\log_\beta^\phi - 1)} + \beta^{(1 - \log_\beta^\phi)} - 2]}{(1 - \log_\beta^\phi)^2} \right\} \frac{k_f \log_\beta^\phi}{(1 - \beta^{\log_\beta^\phi})} D_{av}^{-1}(\Delta T) \tag{2.55}$$

Forster and Greif (1959) and Lin (1988) assumed that the bubble ruptured when it grew into hemisphere and its radius reached R_b (the bubble departure radius) for subcooled boiling in base fluids. For the heat transfer of subcooled boiling, based on the work by Forster and Greif (1959), Lin (1988) derived the heat flux (q_{sb}) as

$$q_{sb} = c_p \rho_f \left(\frac{2}{3} \pi R_b^3 \right) \left(\frac{T_w + T_l}{2} - T_l \right) f N_c \tag{2.56}$$

where
　　R_b is the bubble departure radius
　　f is the bubble departure frequency
　　N_c is the number of active cavity per unit area of heated surfaces

It is well known that subcooled pool boiling heat transfer is characterized by the formation of vapor bubbles that nucleate, grow, and subsequently detach from the locations of active cavities. So it is important to characterize the distribution of active cavities for subcooled pool boiling. Similarly, the number of active cavities whose sizes are within the infinitesimal range from D_c to $D_c + dD_c$ can be expressed as (Yu and Cheng 2002b; Xiao and Yu 2007a,b)

$$-dN_c = d_{fc} D_{c,max}^{d_{fc}} D_c^{-(d_{fc}+1)} dD_c \tag{2.57}$$

where $dD_c > 0$ and $-dN_c > 0$, which means that the number of active cavities decreases with the increase in the diameter of active cavities.

Since the size distribution of active cavities is found to be fractal (Yu and Cheng 2002b; Xiao and Yu 2007a,b), a fractal analytical model for base fluids in subcooled pool boiling from $D_{c,min}$ to $D_{c,max}$ can be obtained by modifying Equation 2.56 as

$$q_{sb} = \int dq_{sb} = \int_{D_{c,min}}^{D_{c,max}} c_p \rho_f \left(\frac{2}{3} \pi R_b^3 \right) \left(\frac{T_w + T_l}{2} - T_l \right) f(-dN_c)$$

$$= \frac{\pi}{3} c_p \rho_f (\Delta T + \Delta T_{sub}) \int_{D_{c,min}}^{D_{c,max}} R_b^3 f(-dN_c) \tag{2.58}$$

Equation 2.58 can be integrated if f and R_b are expressed in terms of D_c. The volume V_b of single bubble at departure is given by Van der Geld (1996) as

$$V_b = \frac{\pi D_c^3}{Eo} \tag{2.59a}$$

where Eo is the Eötvös number, which is given by Mori and Baines (2001) as

$$Eo = \frac{g(\rho_f - \rho_g)D_c^2}{\sigma} \tag{2.59b}$$

If Equation 2.59b is substituted into Equation 2.59a, Equation 2.59a can be further modified as

$$V_b = \frac{\pi\sigma}{g(\rho_f - \rho_g)}D_c \tag{2.59c}$$

We note that the volume of single bubble at departure, V_b, is usually expressed as

$$V_b = \frac{4}{3}\pi R_b^3 \tag{2.60}$$

Comparing Equation 2.59c with Equation 2.60, we can obtain R_b^3 as

$$R_b^3 = \frac{3\sigma}{4g(\rho_f - \rho_g)}D_c \tag{2.61}$$

To this end, we note that the bubble departure frequency (f) is usually expressed as

$$f = \frac{1}{t_w + t_g} \tag{2.62}$$

where
 t_w is the bubble waiting time
 t_g is the bubble growth time

In pure liquids, Van Stralen et al. (1975) assumed that the waiting time is related to the growth time by

$$t_w = 3t_g \tag{2.63a}$$

The bubble waiting time is given by (Yu and Cheng 2002b)

$$t_w = \frac{9}{4\pi\alpha}\left[\frac{(T_w - T_l)D_c}{2(T_w - T_s)}\right]^2 \tag{2.63b}$$

As $T_w - T_l = \Delta T_{sub} + \Delta T$ and $T_w - T_s = \Delta T$ for subcooled pool boiling, and Equation 2.63b can be written as

$$t_w = \frac{9D_c^2(\Delta T + \Delta T_{sub})^2}{16\pi\alpha(\Delta T)^2} \tag{2.63c}$$

Substituting Equations 2.63c and 2.63a into Equation 2.62, we can see that the bubble departure frequency, f, is related to the sizes of active cavities as

$$f = \frac{4\pi\alpha(\Delta T)^2}{3(\Delta T + \Delta T_{sub})^2} D_c^{-2} \tag{2.64}$$

From Equations 2.61 and 2.64, we see that both R_b^3 and f are related to the sizes of active cavities (D_c). So Equation 2.58 can now be integrated to give

$$q_{sb} = \frac{\pi}{3} c_p \rho_f (\Delta T + \Delta T_{sub}) \int_{D_{c,min}}^{D_{c,max}} R_b^3 f(-dN_c)$$

$$= \frac{\pi}{3} c_p \rho_f (\Delta T + \Delta T_{sub}) \int_{D_{c,min}}^{D_{c,max}} \left[\frac{3\sigma}{4g(\rho_f - \rho_g)} D_c \right] \left[\frac{4\pi\alpha(\Delta T)^2}{3(\Delta T + \Delta T_{sub})^2} D_c^{-2} \right]$$

$$\times \left(d_{fc} D_{c,max}^{d_{fc}} D_c^{-(d_{fc}+1)} dD_c \right)$$

$$= \frac{\pi^2 \alpha c_p \rho_f \sigma}{3g(\rho_f - \rho_g)} \frac{d_{fc}}{d_{fc}+1} D_{c,max}^{-1} [\gamma^{-(d_{fc}+1)} - 1] \frac{(\Delta T)^2}{\Delta T + \Delta T_{sub}} \tag{2.65}$$

Equation 2.65 denotes that the heat flux (q_{sb}) is a function of wall superheat, subcooling of fluids, the fractal dimension of active cavities, and the physical properties of fluids. It is found that the ratio of minimum to maximum pore size is 0.001, which best fits the experimental data for fractal objects (Feng et al. 2004; Cai et al. 2010; Cai and Sun 2013). In this work, we consider that the active cavities formed on the heated surfaces are analogous to pores in porous media. Besides, if and only if $\gamma < 10^{-2}$, the number and size of active cavities obey the fractal scaling law (Yu and Cheng 2002b; Xiao and Yu 2007a,b). Additionally, Yang et al. (2001) investigated the active cavity density and size for pool boiling using the experimental system of stainless steel and water. The results (Yang et al. 2001) showed that the ratio of minimum active cavity size to maximum one could reach 0.001. Then the ratio $\gamma = 10^{-3}$ is used for subcooled pool boiling heat transfer of fractal nanofluids in this work.

As $1 < d_{fc} < 2$, $-3 < -d_{fc} - 1 < -2$, $\gamma^{-(d_{fc}+1)} \gg 1$, Equation 2.65 can be further reduced to

$$q_{sb} = \frac{\pi^2 \alpha c_p \rho_f \sigma}{3g(\rho_f - \rho_g)} \frac{d_{fc}\gamma^{-d_{fc}}}{d_{fc}+1} D_{c,min}^{-1} \frac{(\Delta T)^2}{\Delta T + \Delta T_{sub}}$$

(2.66)

The minimum diameter of active cavity is given by Griffith and Wallis (1960) as

$$D_{c,min} = \frac{4\sigma T_s}{\rho_g h_{fg} \Delta T}$$

(2.67)

Inserting Equation 2.67 into Equation 2.66, Equation 2.66 can be further modified as

$$q_{sb} = \frac{\pi^2 \alpha c_p \rho_f \rho_g h_{fg}}{12g(\rho_f - \rho_g)T_s} \frac{d_{fc}\gamma^{-d_{fc}}}{d_{fc}+1} \frac{(\Delta T)^3}{\Delta T + \Delta T_{sub}}$$

(2.68)

Inserting Equations 2.55 and 2.68 into Equation 2.54, a fractal analytical model for subcooled pool boiling of nanofluids can be obtained as

$$q_{sn} = q_{t,c} + q_{sb} = \left\{ \frac{3}{\alpha} \sqrt{\frac{2k_B T}{\pi \rho_n}} \frac{[\beta^{(\log_\beta^\phi - 1.5)} - 1]}{(3 - 2\log_\beta^\phi)} \left[\frac{\beta(2 - \log_\beta^\phi)(1 - \beta^{(1-\log_\beta^\phi)})}{(1 - \log_\beta^\phi)} \right]^{3/2} D_{av}^{-1/2} \right.$$

$$\left. + \frac{2\beta(2 - \log_\beta^\phi)[\beta^{(\log_\beta^\phi - 1)} + \beta^{(1-\log_\beta^\phi)} - 2]}{(1 - \log_\beta^\phi)^2} \right\} \frac{k_f \log_\beta^\phi}{(1 - \beta^{\log_\beta^\phi})} D_{av}^{-1}(\Delta T)$$

$$+ \frac{\pi^2 \alpha c_p \rho_f \rho_g h_{fg}}{12g(\rho_f - \rho_g)T_s} \frac{d_{fc}\gamma^{-d_{fc}}}{d_{fc}+1} \frac{(\Delta T)^3}{\Delta T + \Delta T_{sub}}$$

(2.69)

Equation 2.69 indicates that the heat flux of subcooled pool boiling heat transfer in nanofluids is explicitly related to the nanoparticle concentration (ϕ), the average diameter of nanoparticles (D_{av}), the fractal dimension (d_{fc}) of active cavity on the heated surfaces, the wall superheat (ΔT), and the subcooling of fluids (ΔT_{sub}). No additional/new empirical constant is introduced in this model. It is expected that Equation 2.69 has less empirical constants than conventional models, and every parameter in Equation 2.69 has clear physical meaning. It is expected that the present fractal model can reveal the detailed mechanisms of subcooled pool boiling heat transfer of nanofluids.

2.6.2 Analytical Model for Nucleate Pool Boiling Heat Transfer of Nanofluids

It is generally recognized that there are two main mechanisms contributing to nucleate boiling heat transfer in pure water: the bubble generation and departure from nucleation sites on the superheated surface (q_b) and natural convection on inactive nucleation areas of the heated surface ($q_{f,nc}$) (Xiao et al. 2010). However, in nanofluids there are three main mechanisms contributing to nucleate pool boiling heat transfer: heat convection caused by the Brownian motion of nanoparticles ($q_{t,c}$), the bubble generation and departure from nucleation sites on the superheated surface (q_b), and natural convection on inactive nucleation areas of the heated surface ($q_{f,nc}$) from the base fluids. Thus, the total heat flux (q_{np}) of nucleate pool boiling heat transfer of nanofluids can be expressed as (Xiao et al. 2010)

$$q_{np} = q_{t,c} + q_b + q_{f,nc} \tag{2.70}$$

The density of active sites on the heater surface is affected by the interaction of several parameters on the heater and liquid sides, as well as the liquid–solid contact angle. The distribution of available cavities on the heater surface and the liquid–solid contact angle determine which cavities potentially could activate. At the same time, the transport properties of the heater affect the extent of the thermal interaction among the cavities, causing activation and deactivation of individual cavities. It has been shown that the density or number N_c of active nucleate sites on heated surfaces has a great effect on boiling heat transfer. Several studies have been performed on N_c, which give the functional dependence of N_c on q (heat flux) and ΔT. Some of the notable studies in this field are discussed here. Mikic and Rohsenow (1969) were probably the first to relate active nucleation site density to the sizes of the cavities present on the heated surface and expressed the functional dependence of active nucleation site density on cavity for commercial surfaces. Yang and Kim (1988) made the first attempt to predict quantitatively the active nucleation sites from the knowledge of the size and cone angle distribution of cavities that are actually present on the surface. Using a scanning electron microscope and a differential inference contrast microscope, they established the dependence of the nucleation site density on the characteristics of a boiling surface with the aid of statistical analysis approach. Kocamustafaogullari and Ishii (1988) developed a relation for active nucleation site density in pool boiling. Their correlation expressed active nucleation site density in dimensionless form as a function of dimensionless minimum cavity size and density ratio. Paul and Abdel-Khalik (1983) conducted their experiments on the pool boiling of saturated water at 1 atm along an electrically heated horizontal platinum wire. Using high-speed photography, they measured active nucleation site density and bubble departure diameter up to 70%. Wang and Dhir (1993) were probably the first to perform a systematic study of the effect of contact angle on the density of active nucleation sites. The corrected cavity size D_c was related to the wall superheat for nucleation. It was found that there was a strong influence of wettability on active nucleation site density. Hibiki and Ishii (Hibiki and Ishii 2003) analyzed the effect of the heater surface on the active nucleation site density in boiling systems. It is shown that the active nucleation site density is a function of the cavity size and the contact angle (2003). The nucleate pool boiling heat flux caused by the bubble generation and departure from nucleation sites on the superheated surface as (Xiao and Yu 2007a)

$$q_b = h_{fg}\rho_g V_b f N_c \tag{2.71}$$

So a fractal model for nucleating pool boiling heat flux caused by the bubble generation and departure from nucleation sites on the superheated surface from the minimum site to the maximum site can be obtained by modifying Equation 2.15 as

$$q_b = \int dq_b = \int_{D_{c,\min}}^{D_{c,\max}} h_{fg}\rho_g V_b f(-dN_c) \tag{2.72}$$

$-dN_c$ is given by Equation 2.57 and V_b is given by Equations 2.59c and 2.60 hence, the bubble departure diameter can be obtained from Equations 2.59c and 2.60 as

$$D_b = (1.5\sigma)^{1/3}[g(\rho_f - \rho_g)]^{-1/3} D_c^{1/3} \tag{2.73}$$

The relation between the bubble departure diameter and the bubble departure frequency is given by Rohsenow (1973) as

$$D_b f^2 = 1.32g \tag{2.74}$$

If Equation 2.73 is substituted into Equation 2.74, the bubble departure frequency can be obtained from Equation 2.74 as

$$f = \sqrt{1.32g}\,(1.5\sigma)^{-1/6}[g(\rho_f - \rho_g)]^{1/6} D_c^{-1/6} \tag{2.75}$$

Equation 2.75 indicates that the larger the diameter of nucleation site, the lower the bubble departure frequency. This is consistent with the physical phenomena. This can also mean that the larger diameter of nucleation site causes the larger diameter of bubble at departure. And the larger diameter of bubble at departure requires longer bubble growth time, which leads to lower bubble departure frequency.

From Equations 2.59c and 2.75, it can be seen that both the volume of single bubble at departure and the bubble departure frequency are related to the sizes of nucleation site. So, Equation 2.72 can now be integrated to give

$$q_b = \int dq_b = \int_{D_{c,min}}^{D_{c,max}} h_{fg}\rho_g V_b f(-dN_c)$$

$$= \int_{D_{c,min}}^{D_{c,max}} h_{fg}\rho_g \frac{\pi\sigma}{g(\rho_f - \rho_g)} D_c \sqrt{1.32g} \left(\frac{1.5\sigma}{g(\rho_f - \rho_g)}\right)^{-1/6} D_c^{-1/6} d_{fc} D_{c,max}^{d_{fc}} D_c^{-(d_{fc}+1)} dD_c$$

$$= c_b \frac{d_{fc}(\gamma^{-5/6} - \gamma^{-d_{fc}})}{5 - 6d_{fc}} (\Delta T)^{-5/6} \tag{2.76}$$

where $\quad c_b = 6\pi\sqrt{21.12}[h_{fg}\rho_g \sigma^{10} T_s^5/6g^2(\rho_f - \rho_g)^5]^{1/6} d_{fc}(\gamma^{-5/6} - \gamma^{-d_{fc}})/(5 - 6d_{fc})$. Equation 2.76 denotes that nucleating boiling heat flux caused by the bubble generation and departure from nucleation sites on the superheated surface is a function of wall superheat, fractal dimension of nucleation site, and physical properties of fluid. No additional/new empirical constant is introduced in Equation 2.76, and every parameter in the equation has clear physical meaning.

Natural convection on inactive nucleation areas of the heated surface from the base fluids as given by Mikic and Rohsenow (1969) is

$$q_{f, nc} = (1 - K_1 N_c \pi D_b^2) h_{b, c} \Delta T \tag{2.77}$$

where

K_1 is the proportional constant for bubble diameter of influence, which is taken to be 1.8 by Judd and Hwang (1976)

$h_{b,c}$ is given by Equation 2.32

Yu and Cheng (2002b) compared the pore sizes in porous media to the sizes of nucleation sites and obtained the following expression:

$$\chi = \gamma^{d - d_{fc}} \tag{2.78}$$

where χ is volumetric (or area) fraction, $d = 2$ in the two-dimensional space for heated surfaces. Equation 2.78 can be applied to describe the volume (area) fraction of nucleation sites (Yu and Cheng 2002b). So a fractal model for natural convection on inactive nucleation areas of the heated surface from the base fluids can be obtained by modifying Equation 2.78 as

$$q_{f, nc} = (1 - K_1 \chi) h_{b, c} \Delta T = [1 - K_1 \gamma^{2 - d_{fc}}] h_{b, c} \Delta T \tag{2.79}$$

Equation 2.79 denotes that the single-phase heat flux from the base fluids is a function of temperature, fractal dimension of nucleation site, and the single-phase heat transfer coefficient.

Inserting Equations 2.76 and 2.79 into Equation 2.70, a fractal model for nucleate boiling heat transfer of nanofluids is obtained as

$$q_{np} = q_{t,c} + c_b \frac{d_{fc}(\gamma^{-5/6} - \gamma^{-d_{fc}})}{5 - 6d_{fc}}(\Delta T)^{-5/6} + [1 - K_1\gamma^{2-d_{fc}}]h_{b,c}\Delta T \qquad (2.80)$$

2.6.3 Results and Discussions

Heat transfer characteristics of $CaCO_3$ and Cu nanofluids with and without acoustic cavitation were investigated experimentally (Zhou 2004; Zhou and Liu 2004). The effects of such factors as acoustical parameters, nanoparticle concentration, and fluids subcooling on heat transfer enhancement around a heated horizontal copper tube were discussed in detail (Zhou 2004; Zhou and Liu 2004). Figure 2.18 compares the heat flux of subcooled pool boiling heat transfer from the present model calculated using Equation 2.69 and that from reported experiments (Zhou 2004) at $D_{av} = 90$ nm, $\phi = 0.267$ g/L, and $\Delta T_{sub} = 25$ K for $CaCO_3$ nanofluids. It is seen that there is a fair agreement between the model prediction and the experimental data. The solid line in Figure 2.18 represents the predictions by the present model. Figure 2.15 shows a comparison of the heat flux of subcooled pool boiling heat transfer predicted by the present model (by Equation 2.69) and that from reported experiments (Zhou 2004) at $D_{av} = 80$ nm, $\phi = 0.267$ g/L, and $\Delta T_{sub} = 14$ K for Cu nanofluids. Again, a fair agreement between the model predictions and the experimental data is found. In Figures 2.18 and 2.19,

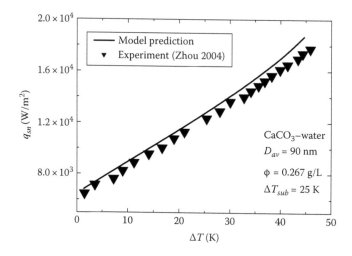

FIGURE 2.18
A comparison between the present model predictions and the experimental data at $D_{av} = 90$ nm, $\phi = 0.267$ g/L, and $\Delta T_{sub} = 25$ K for $CaCO_3$ nanofluids.

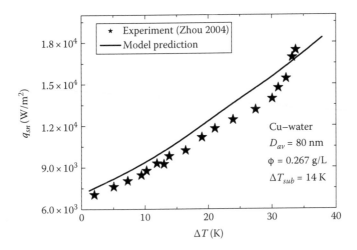

FIGURE 2.19
A comparison between the present model predictions and the experimental data at $D_{av} = 80$ nm, $\phi = 0.267$ g/L, and $\Delta T_{sub} = 14$ K for Cu nanofluids.

the obvious deviations between theoretical and experimental data spotted at large ΔT can be observed. Generally, there are some normal deviations between model predictions and experimental data. Besides, there are many parameters such as subcooling of fluids, nanoparticle concentrations, size of nanoparticles, surface roughness, surface orientation, and contact angle, which can affect the heat transfer of nanofluids. However, the subcooling of fluids and nanoparticle concentrations are same in the literature (Zhou 2004). The experimental data (Zhou 2004) were different when sound source distances between the vibrator head and the central horizontal plane changed.

Das et al. (2003a) conducted an experimental study of pool boiling characteristics of water–Al_2O_3 nanofluid for average nanoparticle size 38 nm and different nanoparticle volume fractions under atmospheric conditions. The heat flux ranged from 2×10^4 to 1.2×10^5 W/m² (Das et al. 2003a). Since the particles under consideration are one to two orders of magnitude smaller than the surface roughness, it was concluded that the change of surface characteristics during boiling due to trapped particles on the surface is the cause of the shift of the boiling characteristics in the negative direction (Das et al. 2003a). Figure 2.20 compares the model predictions with the experimental data by Das et al. (2003a) for average nanoparticle size 38 nm and nanoparticle volume fraction of $\phi = 0.1\%$ and $\phi = 2\%$. It is seen that there is a fair agreement between the model predictions and the experimental data. It is observed that the addition of alumina nanoparticle volume fraction of suspension causes the water boiling curve to shift to the right; that is, it leads to the decrease of pool nucleate boiling heat transfer as shown

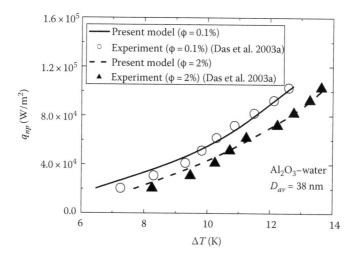

FIGURE 2.20
A comparison between the model predictions and experimental data for nanoparticle diameter 38 nm with two different volume fractions $\phi = 0.1\%$ and $\phi = 2\%$.

in Figure 2.20. This shows that the pool boiling heat transfer coefficient is decreased by increasing the particle concentration, which is consistent with several experimental observations (Das et al. 2003a; Bang and Chang 2005), and appears inconsistent with the increasing thermal conductivity of nanofluids. The experimental results (Das et al. 2003a) indicate that the nanoparticles have pronounced and significant influence on the boiling process, deteriorating the boiling characteristics of the fluids. It has been observed that with the increase in particle concentration, the degradation in boiling performance takes place, and this in turn increases the heating surface temperature (Das et al. 2003a). Boiling heat transfer characteristics of nanofluids with nanoparticles suspended in water were studied using different volume concentrations of alumina nanoparticles (Bang and Chang 2005). This results (Bang and Chang 2005) show that the pool boiling heat transfer coefficient is decreased by increasing particle concentration, which appears inconsistent with the increasing thermal conductivity of nanofluids. Pool boiling heat transfer coefficients and phenomena of nanofluids were compared with those of pure water, which were acquired on a smooth horizontal flat surface (roughness of a few ten nanometers) (Bang and Chang 2005). A flow pattern characterized by a vapor mushroom in high heat flux boiling phenomena was observed in both pure water and nanofluids (Bang and Chang 2005). It is shown that the nanoparticles reduce the number of active nucleation sites with variation of surface roughness values in nucleate boiling heat transfer. Roughness change causes a kind of fouling effect with poor thermal conduction in single-phase heat transfer (Bang and Chang 2005). Comparison between the experimental data and the Rhosenow correlation showed that

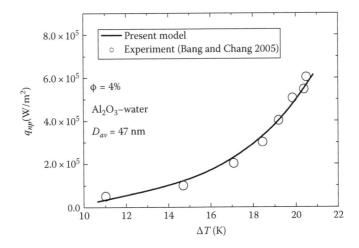

FIGURE 2.21

A comparison between the model predictions and experimental data for nanoparticle diameter 47 nm with volume fractions $\phi = 4\%$.

the correlation could potentially predict the performance with an appropriate modified liquid-surface combination factor and changed the physical properties of the base liquid (Bang and Chang 2005). Boiling heat transfer characteristics of nanofluids were studied using volume fractions $\phi = 4\%$ for an average nanoparticle size of 47 nm in alumina nanoparticle/water fluids (Bang and Chang 2005). Fair agreement is also observed between the experimental and predicted values as shown in Figure 2.21. The nucleate pool boiling heat transfer experiments of water–SiO$_2$ on a plain heated copper surface were carried out under atmospheric pressure (Liu and Liao 2008). The average nanoparticle size was found to be 35 nm with volume fractions $\phi = 0.5\%$ (Liu and Liao 2008). The pool nucleate boiling heat transfer experiments of water-based and alcohol-based nanofluids and nanoparticle suspensions on the plain heated copper surface were carried out. The study was focused on the sorption and agglutination phenomenon of nanofluids on a heated surface. For water-based nanosuspensions, a sorption layer is formed on the heating surface during pool boiling. After the sorption layer is cleaned by water jet, there still exists a very thin coating layer on the surface (Liu and Liao 2008). The surface roughness and the contact angle are decreased due to the formation of the sorption layer (Liu and Liao 2008). For water-based nanofluids, a solid agglutination layer is formed on the heating surface during pool boiling when the wall temperature is over about 112°C (Liu and Liao 2008). Since the thickness of the layer changes irregularly due to the irregular drop of agglomerates from the surface, the wall temperature also changes irregularly, and, hence, steady temperature measurements could not be carried out (Liu and Liao 2008). Figure 2.22

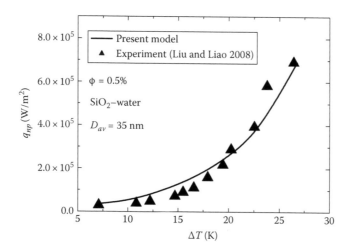

FIGURE 2.22

A comparison between the model predictions and experimental data for nanoparticle diameter 35 nm with volume fractions $\phi = 0.5\%$.

compares our model predictions with experimental data in water–SiO_2 for the average nanoparticle size 35 nm (Liu and Liao 2008). A fair agreement between the model predictions and the experimental data is again found. Figures 2.20 through 2.22 show that the natural convection stage continues relatively longer in the case of nanofluids, which are qualitatively consistent with the other observations (Das et al. 2003a; Bang and Chang 2005; Liu and Liao 2008).

2.6.4 Summary and Conclusions

In this work, based on the fractal distribution of nanoparticles and active cavities on the heated surfaces, we propose two novel fractal models for pool boiling heat transfer of nanofluids, including subcooled pool boiling and nucleate pool boiling. The formula for calculating the heat flux of pool boiling heat transfer is obtained considering the effect of nanoparticles moving in fluids. The models show the dependences of the heat flux on nanoparticle size and the nanoparticle volume fraction of suspension, the fractal dimension of nanoparticle and active cavity, temperature of nanofluids, and properties of fluids. The fractal model predictions show that the natural convection stage continues relatively longer in the case of nanofluids. The addition of nanoparticles causes decrease of the nucleate pool boiling heat transfer. The nucleate pool boiling heat transfer coefficient is decreased by the increasing particle concentration. An excellent agreement between the proposed model predictions and experimental data is found. The validity of the fractal model for pool boiling heat transfer is thus verified.

Nomenclature

A	Surface area of a nanoparticle
A_h	Area of heating surface
A_t	Total surface area of all nanoparticles
C	Empirical constant
c_p	Specific heat at constant pressure
d	Euclidean dimension
D	Diameter of nanoparticle
D_{av}	Average diameter of nanoparticles
D_b	Bubble departure diameter
$D_{c,max}$	Maximum diameter of active cavity
$D_{c,min}$	Minimum diameter of active cavity
d_f	Fractal dimension of nanoparticles
D_f	Diameter of liquid molecule
d_{fc}	Fractal dimension of active cavities
D_{max}	Maximum diameter of nanoparticle
D_{min}	Minimum diameter of nanoparticle
f	Bubble departure frequency
g	Acceleration due to gravity
h	Average heat transfer coefficient caused by the Brownian motion of nanoparticles
$h_{b,c}$	Average heat transfer coefficient by natural convection from the base fluids
h_{fg}	Latent heat of vaporization
k_B	Boltzmann constant
k_c	Thermal conductivity by heat convection caused by the Brownian motion of nanoparticles
k_{eff}	Enhanced effective thermal conductivity of nanofluids
k_{eff}^+	Dimensionless effective thermal conductivity of nanofluids
k_f	Thermal conductivity of base fluids
k_n	Thermal conductivity of nanoparticles
k_s	Thermal conductivity by stationary nanoparticles in liquids
m	Mass of a nanoparticle
N	Number of nanoparticles
N_c	Number of active cavity
N_t	Total number of nanoparticles
Nu	Nusselt number
Pr	Prandtl number
q_b	Heat flux from bubbles
$q_{b,c}$	Heat flux by natural convection from the base fluids
$q_{b,CHF}$	Critical heat flux from base fluids
Q_c	Total heat transfer by convection for all nanoparticles moving in fluids

Q_{D_i}	Heat transferred by convection for a single nanoparticle moving in fluids
$q_{N,CHF}$	Critical heat flux from nanofluids
q_{sb}	Heat flux from the base fluids in subcooled pool boiling
q_{sn}	Heat flux from the nanofluids in subcooled pool boiling
$q_{t,nc}$	Total heat flux of nanofluids by convection
$q_{t,c}$	Total heat flux caused by all nanoparticles moving in fluids
Ra	Rayleigh number
R_b	Bubble departure radius
Re	Reynolds number
T_f	Temperature of fluids
t_g	Bubble growth time
T_l	Bulk temperature of fluids
T_p	Temperature of nanoparticles
T_s	Saturation temperature of fluids
t_w	Bubble waiting time
T_w	Wall temperature
ΔT	Difference of temperature or wall superheat
ΔT_{sub}	Subcooling of fluids
u	Velocity of nanoparticle
V_b	Volume of single bubble at departure

Greek Letters

ϕ	Nanoparticle concentration
σ	Surface tension of liquids
υ	Kinematic viscosity of fluids
α	Thermal diffusivity of fluids
β	Ratio of D_{min}/D_{max}
θ	Contact angle
ρ_f	Density of base liquids
ρ_g	Vapor density
ρ_n	Density of nanoparticles
λ	Length scale
γ_1	Volumetric thermal expansion coefficient
γ	Ratio of $D_{c,min}/D_{c,max}$
δ	Laminar boundary layer
δ_T	Thickness of thermal boundary layer of heat convection caused by Brownian motion
Ja^*	Jakob number

Subindexes

av	Average
B	Bubble
c	Cavity
f	Fluids
max	Maximum
min	Minimum
n	Nanoparticle
nc	Natural convection
s	Saturation
t	Total

Acknowledgments

This work was supported by the National Natural Science Foundation of China (Grant No. 11102100), the Scientific Research Foundation for Middle-aged and Young Teachers of Educational Department of Fujian Province of China (Grant No. JA14285), and the Program for Young Top-notch Innovative Talents of Fujian Province of China.

References

Acrivos, A., T.D. Taylor. Heat and mass transfer from single spheres in stokes flow. *Physics of Fluids* 5(4) (1962): 387–394.

Bang, I.C., S.H. Chang. Boiling heat transfer performance and phenomena of Al_2O_3–water nano-fluids from a plain surface in a pool. *International Journal of Heat and Mass Transfer* 48(12) (2005): 2407–2419.

Cai, J.C., X.Y. Hu, D.C. Standnes, L.J. You. An analytical model for spontaneous imbibition in fractal porous media including gravity. *Colloids and Surfaces A: Physicochemical and Engineering Aspects* 414 (2012): 228–233.

Cai, J.C., S.Y. Sun. Fractal analysis of fracture increasing spontaneous imbibition in porous media with gas-saturated. *International Journal of Modern Physics C* 24(8) (2013): 1350056.

Cai, J.C., B.M. Yu. Prediction of maximum pore size of porous media based on fractal geometry. *Fractals* 18(4) (2010): 417–423.

Cai, J.C., B.M. Yu. A discussion of the effect of tortuosity on the capillary imbibition in porous media. *Transport in Porous Media* 89(2) (2011): 251–263.

Cai, J.C., B.M. Yu, M.Q. Zou, L. Luo. Fractal characterization of spontaneous co-current imbibition in porous media. *Energy & Fuels* 24(3) (2010): 1860–1867.

Das, S.K., N. Putra, W. Roetzel. Pool boiling characteristics of nano-fluids. *International Journal of Heat and Mass Transfer* 46(5) (2003a): 851–852.

Das, S.K., N. Putra, P. Thiesen, W. Roetzel. Temperature dependence of thermal conductivity enhancement for nanofluids. *ASME Journal of Heat Transfer* 125(4) (2003b): 567–574.

Feng, Y.J., B.M. Yu, M.Q. Zou, D.M. Zhang. A generalized model for the effective thermal conductivity of porous media based on self-similarity. *Journal of Physics D: Applied Physics* 37(21) (2004): 3030–3040.

Forster, D.E., F. Greif. Heat transfer to a boiling liquid mechanism and correlations. *ASME Journal of Heat Transfer* 81 (1959): 43–53.

Griffith, P., J.D. Wallis. The role of surface conditions in nucleate boiling. *Chemical Engineering Progress Symposium* 56(30) (1960): 49–63.

Gupta, A., R. Kumar. Role of Brownian motion on the thermal conductivity enhancement of nanofluids. *Applied Physics Letters* 91(22) (2007): 223102.

Han, C.Y., P. Griffith. The mechanism of heat transfer in nucleate pool boiling. *International Journal of Heat and Mass Transfer* 8(6) (1965): 887–904.

He, Y.R., Y. Jin, H.S. Chen, Y.L. Ding. Heat transfer and flow behaviour of aqueous suspensions of TiO_2 nanoparticles (nanofluids) flowing upward through a vertical pipe. *International Journal of Heat and Mass Transfer* 50(11–12) (2007): 2272–2281.

Hibiki, T., M. Ishii. Active nucleation site density in boiling systems. *International Journal of Heat and Mass Transfer* 46(14) (2003): 2587–2601.

Hong, K.S., T.K. Hong, H.S. Yang. Thermal conductivity of Fe nanofluids depending on the cluster size of nanoparticles. *Applied Physics Letters* 88(3) (2006): 031901.

Hsu, Y.Y. On the size range of active nucleation cavities on a heating surface. *ASME Journal of Heat Transfer* 84(3) (1962): 207–215.

Jang, S.P., S.U.S. Choi. Role of Brownian motion in the enhanced thermal conductivity of nanofluids. *Applied Physics Letters* 84(21) (2004): 4316–4318.

Judd, R.L., K.S. Hwang. A comprehensive model for nucleate pool boiling heat transfer including microlayer evaporation. *ASME Journal of Heat Transfer* 98(4) (1976): 623–629.

Kim, H., H.S. Ahn, M.H. Kim. On the mechanism of pool boiling critical heat flux enhancement in nanofluids. *ASME Journal of Heat Transfer* 132(6) (2010): 061501.

Kim, H.D., J. Kim, M.H. Kim. Experimental studies on CHF characteristics of nanofluids at pool boiling. *International Journal of Multiphase Flow* 33(7) (2007a): 691–706.

Kim, S.J., I.C. Bang, J. Buongiorno, L.W. Hu. Surface wettability change during pool boiling of nanofluids and its effect on critical heat flux. *International Journal of Heat and Mass Transfer* 50(19–20) (2007b): 4105–4106.

Kocamustafaogullari, G., M. Ishii. Interfacial area and nucleation site density in boiling systems. *International Journal of Heat and Mass Transfer* 31(6) (1988): 1127–1135.

Koo, J., C. Kleinstreuer. A new thermal conductivity model for nanofluids. *Journal of Nanoparticle Research* 6 (2004): 577–588.

Lin, R.T. *Boiling Heat Transfer*, 1st edn. Beijing, China: Science Press, 1988.

Liu, Z.H., L. Liao. Sorption and agglutination phenomenon of nanofluids on a plain heating surface during pool boiling. *International Journal of Heat and Mass Transfer* 51(9–10) (2008): 2593–2602.

Mandelbrot, B.B. *The Fractal Geometry of Nature.* New York: W. H. Freeman, 1982.

Masuda, H., A. Ebata, K. Teramae, N. Hishinuma. Alteration of thermal conductivity and viscosity of liquid by dispersing ultra-fine particles (dispersions of $Y-A1_2O_3$, SiO, and TiO_2 ultra-fine particles). *Netsu Bussei (Japan)* 4 (1993): 227–233.

Maxwell, J.C. 1873. *Treatise on Electricity and Magnetism.* New York: Oxford.

Mikic, B.B., W.M. Rohsenow. A new correlation of pool boiling data including the effect of heating surface characteristics. *ASME Journal of Heat Transfer* 91(2) (1969): 245–250.

Mori, B.K., W.D. Baines. Bubble departure from cavities. *International Journal of Heat and Mass Transfer* 44(4) (2001): 771–783.

Paul, D.D., S.I. Abdel-Khalik. A statistical analysis of saturated nucleate boiling along a heat wire. *International Journal of Heat and Mass Transfer* 29(4) (1983): 509–519.

Prasher, R., P. Bhattacharya, P.E. Phelan. Thermal conductivity of nanoscale colloidal solutions (nanofluids). *Physical Review Letters* 94(2) (2005): 025901.

Rohsenow, W.M. *Handbook of Heat Transfer.* New York: McGraw-Hill, 1973.

Shima, P.D., J. Philip, B. Raj. Role of microconvection induced by Brownian motion of nanoparticles in the enhanced thermal conductivity of stable nanofluids. *Applied Physics Letters* 94(22) (2009): 223101.

Van der Geld, C.W.M. Bubble detachment criteria: Some criticism of 'Das Abreissen von Dampfblasen an festen Heizflächen'. *International Journal of Heat and Mass Transfer* 39(3) (1996): 653–657.

Van Stralen, S.J.D., M.S. Sohan, R. Cole, W.M. Sluyter. Bubble growth rates in pure and binary systems: Combined effect of relaxation and evaporation microlayers. *International Journal of Heat and Mass Transfer* 18(3) (1975): 453–467.

Wang, B.X., L.P. Zhou, X.F. Peng. A fractal model for predicting the effective thermal conductivity of liquid with suspension of nanoparticles. *International Journal of Heat and Mass Transfer* 46(14) (2003): 2665–2672.

Wang, C.H., V.K. Dhir. Effect of surface wettability on active nucleation site density during pool boiling of water on a vertical surface. *ASME Journal of Heat Transfer* 115(3) (1993): 659–669.

Xiao, B.Q. Prediction of heat transfer of nanofluid on critical heat flux based on fractal geometry. *Chinese Physics B* 22(1) (2013): 014402.

Xiao, B.Q., G.P Jiang, L.X. Chen. A fractal study for nucleate pool boiling heat transfer of nanofluids. *Science China Physics, Mechanics & Astronomy* 53(1) (2010): 30–37.

Xiao, B.Q., G.P. Jiang, Y. Yang, D.M. Zheng. Prediction of convective heat transfer of nanofluids based on fractal-Monte Carlo simulations. *International Journal of Modern Physics C* 24(1) (2013a): 1250090.

Xiao, B.Q., Y. Yang, L.X. Chen. Developing a novel form of thermal conductivity of nanofluids with Brownian motion effect by means of fractal geometry. *Powder Technology* 239 (2013b): 409–414.

Xiao, B.Q., Y. Yang, X.F. Xu. Subcooled pool boiling heat transfer in fractal nanofluids: A novel analytical model. *Chinese Physics B* 23(2) (2014): 026601.

Xiao, B.Q., B.M. Yu. A fractal model for critical heat flux in pool boiling. *International Journal of Thermal Sciences* 46(5) (2007a): 426–433.

Xiao, B.Q., B.M. Yu. A fractal analysis of subcooled flow boiling heat transfer. *International Journal of Multiphase Flow* 33(10) (2007b): 1126–1139.

Xiao, B.Q., B.M. Yu, Z.C. Wang, L.X. Chen. A fractal model for heat transfer of nanofluids by convection in a pool. *Physics Letters A* 373(45) (2009): 4178–4181.

Xie, H.Q., J.C. Wang, T.G. Xi, Y. Liu, F. Ai, Q.R. Wu. Thermal conductivity enhancement of suspensions containing nanosized alumina particles. *Journal of Applied Physics* 91(7) (2002): 4568–4572.

Yang, S.R., R.H. Kim. A mathematical model of pool boiling nucleation site density in terms of surface characteristics. *International Journal of Heat and Mass Transfer* 31(6) (1988): 1127–1135.

Yang, S.R., Z.M. Xu, J.W. Wang, X.T. Zhao. On the fractal description of active nucleation site density for pool boiling. *International Journal of Heat and Mass Transfer* 44(14) (2001): 2783–2786.

Yu, B.M., P. Cheng. A fractal permeability model for bi-dispersed porous media. *International Journal of Heat and Mass Transfer* 45(14) (2002a): 2983–2993.

Yu, B.M., P. Cheng. A fractal model for nucleate pool boiling heat transfer. *ASME Journal of Heat Transfer* 124(6) (2002b): 1117–1124.

Yu, C.J., A.G. Richter, A. Datta, M.K. Durbin. Observation of molecular layering in thin liquid films using x-ray reflectivity. *Physical Review Letters* 82(11) (1999): 2326–2329.

Zhou, D.W. Heat transfer enhancement of copper nanofluid with acoustic cavitation. *International Journal of Heat and Mass Transfer* 47(14–16) (2004): 3109–3117.

Zhou, D.W., D.Y. Liu. Heat transfer characteristics of nanofluids in an acoustic cavitation field. *Heat Transfer Engineering* 25(6) (2004): 54–61.

3

Thermal Conductivity Enhancement in Nanofluids Measured with a Hot-Wire Calorimeter

Catalina Vélez, José M. Ortiz de Zárate, and Mohamed Khayet

CONTENTS

3.1 Introduction 79
3.2 Theory and Method 81
3.3 Experimental Setup 84
3.4 Preparation of the Nanofluids 88
3.5 Experimental Results 90
    3.5.1 Statistics 90
    3.5.2 Nanofluids 91
3.6 Discussion 93
    3.6.1 Comparison with Available Experimental Data 93
    3.6.2 Effective Medium Theoretical Models 95
Acknowledgments 98
References 98

3.1 Introduction

The last decade has shown a growing interest in the measurement and modelization of the transport properties of nanofluids, that is, dispersions of nanoparticles in the bulk of simple fluids. Even a dedicated *Journal of Nanofluids* has been recently established (2012). Since the topic is well covered by excellent and updated reviews [1–4], this introductory section to our chapter will be brief and centered more in the particular features of our contribution than in the general aspects of the problem for which we refer the interested reader to Refs. [1–4]. We just mention that, historically, this nanofluid effort was triggered by Eastman et al. [5], who reported large effective thermal conductivity enhancements when a small amount of nanoparticles was added to a liquid. For this reason, among the various nanofluid properties analyzed, the thermal conductivity (λ) has received the most attention. Only more recently, other transport properties, viscosity in particular [6–9], have also been investigated. It is obvious that λ alone is not enough to assess the practical usefulness of nanofluids in industry or laboratory applications, and a full characterization of their thermophysical properties is required.

Besides questions related to actual applications, from a fundamental point of view, maybe the most interesting issue associated with nanofluids is whether the observed effective thermal conductivity enhancements ($\Delta\lambda$) are *anomalous* or can be explained by standard theoretical models for the thermal conductivity of solid suspensions, like the classical Maxwell model [10] as adapted by Hamilton and Crosser [11]. This has been a controversial issue [12]. However, the current consensus is that in most cases the observed enhancements can be simply explained by accounting for the volume of the system that is occupied by solids having an intrinsically larger λ (effective medium theories). Indeed, in the year 2009, a large collaborative effort [13] measured the thermal conductivity of the same nanofluid samples in different laboratories and with different techniques, concluding that the measured $\Delta\lambda$ are, with a high probability, not anomalous. In any case, more experimental work will be necessary during the coming years to finally settle the question.

In this contribution, we review the measurements of the effective thermal conductivity of nanofluids performed at the Applied Physics I Laboratory of the Complutense University (Madrid, Spain) during last years. For these measurements, we have employed a customized hot-wire calorimeter developed by our research group [14]. For the sake of completeness, experimental results are reviewed, including both previously published data and those reported in the present chapter for the first time.

The main advantage of our measurement method is that we have full control of the experimental conditions, particularly of temperature and heating intensity. This flexibility is particularly desirable when comparing with more compact measurement apparatus, like the commonly used Decagon Devices KD2© series. Our advantage is that, depending on the sample under testing, we can choose the best working electrical current intensity (or average over several intensities) to obtain more reliable λ values. Similarly, we can freely vary the duration of the heating runs to avoid any problem related to convection in the sample.

Of course, the flexibility of our experimental setup has some associated drawbacks. In particular, data analysis is especially long and complicated. First, we have to pay attention to the theory of the hot-wire and be sure that the theoretical approximations performed are good enough to explain the experimental curves. In addition, for each sample and each temperature, we typically performed several hundreds of individual measurements, and one has to reduce all this information to a single number. For these reasons, we will dedicate space in the present chapter to report in some detail the theory of hot-wire calorimetry and the statistics analysis we have performed for data reduction.

In consequence, we have organized the material to be presented as follows. In Section 3.2, we summarize the theory behind the hot-wire calorimetry and review the experimental method. Then, in Section 3.3, we briefly describe the experimental setup. Subsequently, after reporting on the nanofluid preparation in Section 3.4, we review the experimental results obtained by us for the effective thermal conductivity of the different samples in Section 3.5, including an explanation of the associated statistical analysis. Finally, in Section 3.6, we discuss our results and compare them with other experimental values published in refereed scientific literature, as well as with various theoretical models proposed for the prediction of the effective thermal conductivity of nanofluids.

3.2 Theory and Method

Values for the effective thermal conductivity of the nanofluids were obtained by the transient hot-wire technique. Nowadays, this technique is accepted as the most precise and reliable method to measure the thermal conductivity of fluids over a wide range of temperatures and pressures. A nice historical account of the development of this technique, as well as a current state-of-the-art review, has been recently published by Assael et al. [15]. The measurements can deliver absolute values (do not require comparison with any standard), and only knowledge of the applied intensity, the wire geometry, and its electrical resistance is required. For the best experiments, the accuracy of the results is estimated to be as good as ±0.5% [15–17].

When an electrical current I is circulated through a straight wire that is surrounded by a fluid, due to Joule's effect, the wire acts as a heat source. Hence, the temperature of the fluid (and of the wire itself) increases. While the fluid surrounding the wire is quiescent (no convection), its temperature can be obtained by solving the heat equation only [18] (further assuming that the fluid is electrically not conducting and/or not polarizable). If one assumes that the wire has infinite length and chooses the vertical z-axis along the wire, the problem will have cylindrical symmetry. If one further assumes that the wire has zero radius, it can be considered as a linear heat source with intensity $\dot{Q}$ (watts per unit length) given by Joule's effect. With these assumptions, the solution $T(r,t)$ of the heat equation can be readily expressed by using the corresponding Green (or source) function in cylindrical coordinates [18]:

$$T(r,t) = T_0 + \frac{1}{4\pi\lambda} \int_0^t \frac{\dot{Q}}{\tau} \exp\left(-\frac{r^2 \rho c_p}{4\lambda\tau}\right) d\tau \tag{3.1}$$

where
  T_0 is the equilibrium temperature when the heating starts
  ρ is the mass density
  c_p is the isobaric specific heat (enthalpy) of the medium

The time origin in Equation 3.1 is when the current starts to circulate through the wire, that is, $\dot{Q} \equiv 0$ for negative time. Equation 3.1 gives the temperature in the medium as a function of time t and distance r to the linear heat source. We next assume that the temperature of the wire is given by Equation 3.1 itself, evaluated at $r = r_0$, where r_0 is the wire radius. For $r < r_0$, since the thermal conductivity of the wire material is much larger than that of the surrounding medium, the temperature will be uniform and equal to $T(r_0,t)$. Hence, the increment of the wire temperature during the experiment will be $\Delta T(t) = T(r_0,t) - T_0$. If one further assumes that the source intensity $\dot{Q}$ is constant, the integral in Equation 3.1 can be evaluated analytically, and one can obtain the temperature increment:

$$\Delta T(t) = \frac{\dot{Q}}{4\pi\lambda} \mathrm{Ei}\left(\frac{r_0^2 \rho c_p}{4\lambda t}\right) \tag{3.2}$$

where $\mathrm{Ei}(x)$ is the exponential integral function [19]. For long time $t \gg r_0^2 \rho c_p / 4\lambda$, the function $\mathrm{Ei}(x)$ can be substituted by its asymptotic expansion at $x \to 0$, namely [19,20]:

$$\Delta T(t) \simeq \frac{I^2 R_0}{4\pi\lambda L}\left[-\gamma + \ln\left(\frac{t}{\beta}\right)\right] \tag{3.3}$$

where γ is Euler's constant ($\gamma = 0.5770$) and the parameter β (units of time) is expressed as

$$\beta = \frac{r_0^2}{4a} \tag{3.4}$$

with $a = \lambda/\rho c_p$ being the thermal diffusivity of the medium surrounding the wire. Furthermore, we have substituted in Equation 3.3 a constant Joule heating, $\dot{Q} = I^2 R_0/L$, with R_0 the wire electrical resistance at $t = 0$ and L the wire finite length. In this substitution, two approximations are implicit: finite-size effects are neglected as well as any self-heating. Finally, we mention that, as clarified later, the long times considered to get Equation 3.3 are, in practice, a few milliseconds.

Next, we consider that the temperature increase in the wire given by Equation 3.3 will cause an increase in its electrical resistance. For metals, a linear variation of resistance with temperature is usually assumed:

$$R(T) = R_0(1 + \alpha \Delta T) \tag{3.5}$$

with α the temperature resistance coefficient of the wire material. Finally, by Ohm's law, one finds that the potential difference between the wire ends will be given by [14]

$$V(t) \simeq IR_0 \left\{ 1 + \alpha \frac{I^2 R_0}{4\pi\lambda L} \left[\ln\left(\frac{t}{\beta}\right) - \gamma \right] \right\} \tag{3.6}$$

Equation 3.6 represents the most simple and direct way of analyzing hot-wire calorimetry experiments. From the slope b of $V(t)$ versus $\ln t$ experimental lines (examples of which are shown later in Figure 3.2), the thermal conductivity of the fluid can be obtained as

$$\lambda = \frac{mI^3 R_0}{4\pi Lb} \tag{3.7}$$

where the quantity $m = \alpha R_0$ represents the slope of the $R(T)$ curve at the initial temperature of a heating run. By using Equation 3.7, absolute λ values can be obtained from known values of the other properties involved.

The accuracy of the λ values obtained from Equation 3.7 is mainly determined by the validity of the theory. The effects of finite length of the wire [20], leaking by radiation [21], compression work [22], nonzero time required to ramp the current, self-heating, etc., are not accounted for in Equations 3.1 through 3.7. All these factors contribute as systematic errors and limit the accuracy of the absolute measurements based on Equation 3.7 to a few percent, which for our particular setups has been estimated as 2.5%–3% [14]. The inclusion of all or some of these unaccounted effects, by using numerical integration methods [23] if needed, can certainly improve the accuracy of the λ values obtained from absolute measurements.

However, our present investigation is focused on the effective λ of nanofluids, and our primary goal is to measure the thermal conductivity enhancement, $\Delta\lambda$, and not the thermal conductivity itself. For this reason, we used in our nanofluid investigation a relative measurement method [24,25], for which improvements of Equation 3.7 for data analyzing and reduction are not so important. The idea of the relative method is to rewrite Equation 3.7 as

$$\lambda = A \frac{I^3 R_0}{b} \tag{3.8}$$

where A is a constant that depends only on the wire characteristics and not on the fluid under test or the temperature. A calibration fluid whose thermal conductivity is known can be used to infer a value for A. If measurements are performed at various temperatures and/or with different fluids, a better estimation of A can be obtained by simultaneous comparison with a set of tabulated $\lambda(T)$. Once A is determined, this value is used to evaluate the thermal conductivity of the test fluid. Initially, parameter A includes the length of the wire and the slope of the $\{R,T\}$ straight line, but it also represents all unknown sources of systematic errors. We estimate that this calibration method improves the accuracy of λ values, reducing the contribution of systematic errors to around 1% [24,25].

As mentioned earlier, this calibration procedure is especially well suited for the study of the thermal conductivity of the nanofluids. Indeed, for nanofluids, one is specifically interested in the so-called thermal conductivity enhancement, $\Delta\lambda$, defined as [26]

$$\Delta\lambda\,(\%) = \frac{\lambda_{\text{nanofluid}} - \lambda_{\text{liquid}}}{\lambda_{\text{liquid}}}\,100 \tag{3.9}$$

From Equation 3.8, it is obvious that the experimental enhancement of the thermal conductivity is insensitive to the calibration constant A. Hence, more accurate values for $\Delta\lambda$ are expected.

Consequently, for our investigation, we performed experimental series not only with nanofluids but also with the base fluids, that is, pure water and pure ethylene glycol. These pure fluids series serve for calibration of the wires and as baselines for evaluating the thermal conductivity enhancements.

For the work reviewed in this chapter, we used three different wires since they typically break after 3–4 weeks of continuous use. Each time that the wire had to be replaced, a new calibration series was performed. We have found that the A values that best represent the known thermal conductivities of water and ethylene glycol differ by a few percent (2%–4%, depending on the wire) from the values expected on the basis of the length of the wire and the slope of the experimental $\{R,T\}$ line. This difference accounts for the various systematic errors and gives an estimation of the accuracy of our measurements.

3.3 Experimental Setup

The experimental setup used in this research is, essentially, the same previously employed in other investigations [14,24,25]. In Figure 3.1, we show a schematic representation of the various components of our setup, which we describe next in some detail.

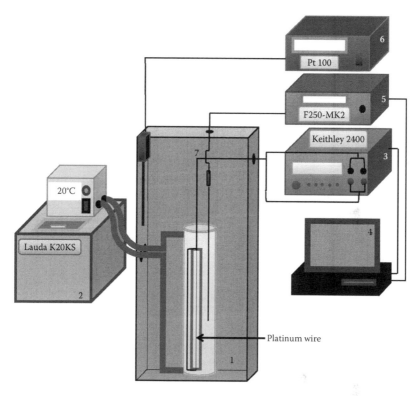

FIGURE 3.1

(See color insert.) Schematic representation of the experimental setup for the measurement of the thermal conductivity. (1) Measurement cell, (2) Lauda K20KS thermostatic circulation bath, (3) Keithley 2400 source meter, (4) computer, (5) ASL F250 reference thermometer, (6) Pt 100 digital thermometer for monitoring chamber temperature, and (7) atmospheric chamber with controlled temperature and humidity (Mytron).

The core of our experimental system is a platinum wire of 50 μm diameter. The length of the wires used in this research was around 21 cm. For good measurements, it is important to keep the wire straight, which is achieved by soldering the two wire ends to tabs in a chemically resistant flat frame cut from a circuit board. Two Teflon isolated cables were then soldered to each one of the tabs, after which the connections were covered by a thermal-resistant epoxi. Furthermore, to avoid electrical contact between the platinum wire and the samples, the whole set (including the wire itself and the frame supporting it) was covered with a Teflon-based industrial coating. The thickness of this coating is less than 1 μm.

A Keithley 2400 source meter, which can act simultaneously as current source and voltage meter, was employed for electrical measurements. This instrument is interfaced to a personal computer; and a software code was written to retrieve the measurement points, fit the data, and obtain the

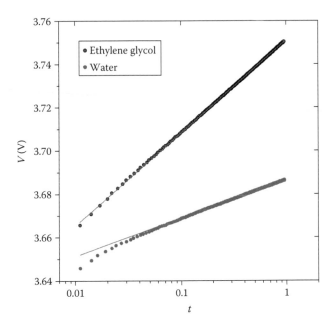

FIGURE 3.2
(See color insert.) Typical individual heating curves for one of the platinum wires used in this research: voltage drop in the wire, V, as a function of time, t. Blue symbols are for the wire immersed in pure ethylene glycol, and red symbols for the wire immersed in pure water. Data are for an initial temperature of 40°C and heating current $I = 260$ mA. Straight lines represent fittings to Equation 3.6 of the points measured after $\simeq 213$ ms.

slope b of the $\{V, \ln t\}$ line (see Figure 3.2 and Equations 3.6 and 3.7). The electrical current is injected through two of the leads connected to the wire ends, while voltage measurements are acquired simultaneously using the other two leads. For each nanofluid and each temperature, various current values, from 260 to 360 mA, were applied. The supported wire was placed vertically inside a double wall glass cell. The liquid under testing is loaded in the inner volume of the glass cell. The cell jacket is connected to a thermostatic bath Lauda K20KS that controls the temperature at which the experiments are performed within ±0.05 K. The whole assembly (glass cell containing the wire and the liquid sample) is then placed inside a controlled atmosphere chamber (temperature and humidity, Mytron). To minimize heat losses and improve stability, the temperature of the chamber is programmed to the same value of the circulation bath. Chamber temperature is controlled within ±0.5 K. Furthermore, a reference platinum resistance thermometer (ASL F250) is placed inside the cell to measure the actual temperature of the liquid under testing. This thermometer is also interfaced to the personal computer controlling the experiment, and the same computer code manages simultaneously the thermometer and the Keithley 2400 source meter. Finally, temperature inside the Mytron atmosphere chamber is monitored with a regular Pt100 thermometer.

In our experiments, a typical heating run lasts for a couple of seconds during which 350 voltage measurements are acquired. Before each heating run, a resistance measurement is performed using the four-wire configuration of the Keithley 2400, and a temperature reading of the reference thermometer is also recorded. The data pairs $\{V_i, \ln(t_i)\}$ acquired during the heating run are then fitted to a straight line, from which a slope b is obtained. To make sure that we are in the asymptotic large time regime of Equation 3.3, only the points acquired after $\simeq 213$ ms are actually used in the fitting procedure. From each heating run, i, values of temperature T_i, electrical resistance $R_{0,i}$, current intensity I_i, and slope b_i are stored in a computer file for further analysis.

As an example of individual heating runs, we show in Figure 3.2 two typical heating curves, where the voltage drop in the platinum wire is displayed as a function of time in a semilogarithmic scale. Data are shown for the wire immersed in water and in ethylene glycol, as indicated. In both cases, the initial temperature was 40°C and the heating current, $I = 280$ mA. The data displayed in Figure 3.2 show that the experimental results are asymptotically well represented by Equation 3.6 and that no convection occurs in the liquid.

One of the assumptions of the theoretical analysis presented in Section 3.2 is that the electrical resistance of the wire, $R(T)$, depends linearly on temperature; see Equation 3.5. As explained earlier, the computer code that controls our experiments stores, just before performing each heating run, the electrical resistance measured with the Keithley 2400, and the corresponding reference temperature measured with the ASL F250 thermometer. Hence, it allows for an easy verification of the linearity of $R(T)$. In addition, frequent analysis of actual $R(T)$ lines, and comparison with $R(T)$ obtained previously for the same wire, has been used as a proxy for wire stability and aging. Indeed, if differences in electrical resistance larger than $\pm 0.5\%$ were detected in these comparisons, the corresponding wire was discharged. In Figure 3.3, we show the electrical resistance of one of the wires used in this research as a function of temperature, $R(T)$. Note that each one of the points shown in Figure 3.3 is an average over hundreds of separate measurements, performed before individual heating runs. We add in Figure 3.3 a straight line that represents a least-squares fitting of all the data points to Equation 3.5. The linearity of $R(T)$ is excellent, and the theory is verified. We further note that the temperature coefficients α obtained from linear fits as the one shown in Figure 3.3 agree reasonably well with tabulated values for the α of platinum. Likewise, Figure 3.3 also shows good wire stability during the several weeks of uninterrupted wire use needed to obtain all the data represented on it.

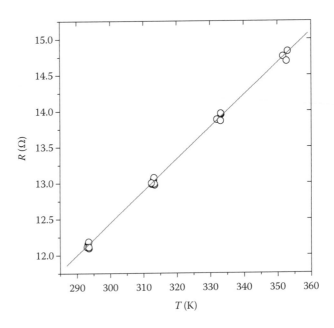

FIGURE 3.3
Electrical resistance of one of the platinum wires used in this research as a function of the temperature recorded inside the measurement cell. Four series of measurements are represented. Two were simultaneous to calibration runs measuring the λ of pure ethylene glycol, while the other two were simultaneous to the measurement of the $\Delta\lambda$ of the aluminum nitride (AlN) nanofluid. The straight line represents a fitting of all the data to Equation 3.5.

3.4 Preparation of the Nanofluids

The liquids employed as base fluids in our investigation were water (W) and ethylene glycol (EG). Water was prepared in our laboratory by double distillation, its electrical resistivity being higher than 15 MΩ cm. The liquid EG was supplied by *Panreac Química S.A.* (Barcelona, Spain), with 99.5% nominal purity, and used without further purification.

In our research, we used commercial nanopowders supplied by *Sigma–Aldrich*. Silica (SiO_2) nanopowder is described by the manufacturer (catalog#: S5505) as having an average particle size of 14 nm and a surface area of 200 ± 25 m^2 g^{-1}. Copper(II) Oxide (CuO) nanopowder is described by the manufacturer (catalog#: 544868) as having a particle size of less than 50 nm and a surface area of 29 m^2 g^{-1}. Finally, AlN nanopowder is described by the manufacturer (catalog#: 593044) as having a particle size less than 100 nm.

Nanofluids were prepared by weighting in a balance having a precision of ± 0.05 g. Eight nanofluids were prepared, three based on water (W1–W3) and five based on EG (EG1–EG5). In Table 3.1, we summarize the characteristics

TABLE 3.1

Nanofluids Investigated in This Research, Including the Nanopowder Concentration in Weight Fraction and the Corresponding Volume Fraction, Estimated by Using Mass Densities of the Bulk Solids and Liquids

Nanofluid	Nanopowder	Particle Size (nm)	Fluid	Particle (Weight Fraction) (%)	Estimated (ϕ) (%)
W1	SiO_2	14	Water	4.8	2.2
W2	CuO	<50	Water	2.4	0.4
W3	CuO	<50	Water	4.9	0.8
EG1	SiO_2	14	Ethylene glycol	2.3	1.2
EG2	SiO_2	14	Ethylene glycol	4.8	2.5
EG3	CuO	<50	Ethylene glycol	2.2	0.4
EG4	CuO	<50	Ethylene glycol	4.6	0.8
EG5	AlN	<100	Ethylene glycol	1.5	0.5

of the nanofluids investigated. The fifth column of Table 3.1 contains the concentration (in weight fraction, w/w) of nanopowder used to prepare each sample. Estimations of the volume fraction ϕ of particles are displayed in the last column. The ϕ values have been obtained from the densities of the liquids and of the bulk solids (at 20°C). These densities were obtained from standard thermodynamic tables. We note that there is a huge difference between the densities of the bulk solids and the corresponding nanopowders. This is most likely caused by strong electrostatic repulsion among the nanoparticles. Once the nanofluids are prepared, and if the particles are indeed well dispersed in the liquid, we think that a correct estimation of ϕ is obtained from the density of the bulk solids, not of the nanopowders.

Dispersions were performed only by physical means without using any chemical additive (surfactant, pH buffer, or any other kind). Mixtures were first strongly stirred mechanically and later subjected to ultrasonics (150 W power) for, at least, half an hour to break up any residual agglomerations. After this procedure, very homogeneous dispersions were obtained. It is well known that one of the most problematic issues when working with nanofluids is sample stability [27]. Nanoparticles tend to sediment at the bottom of the sample containers, causing the physical properties of the bulk nanofluid to change with time. In our case, the measurement of the thermal conductivity of the nanofluid requires temperature cycling and many individual measurements, the whole process lasting around 1 week. In our research, special care has been taken to check sample stability. After finalizing the measurements, we checked the samples and found no visible sedimentation. Furthermore, in some cases, we repeated the temperature cycling of the measurements (without further stirring of the samples), looking for differences in the measured λ that might indicate lack of stability. We found no differences larger than the accuracy of the λ measurements.

We stress that the results reviewed here were all obtained in the first weeks after sample preparation and were checked for reproducibility. In this study, only λ values are presented for nanofluids, showing reproducibility and sample homogeneity during the measurement period.

3.5 Experimental Results

As discussed earlier, the thermal conductivity measurements of the nanofluids have been performed by the calibration method using the corresponding base fluid (water or ethylene glycol) as calibration standard. For two of the wires used in this research, calibration runs were performed with both water and ethylene glycol at the same temperatures for which effective λ of the nanofluids was later measured, namely, temperatures around 20°C, 40°C, and 60°C for water and temperatures around 20°C, 40°C, 60°C, and 80°C for ethylene glycol. Note that all temperatures reported in what follows are averaged over the actual readings inside the fluid cell, which slightly differ from temperatures programmed in the thermostat. For one of the wires used in our nanofluid investigation, calibration runs were performed only with ethylene glycol. This particular wire was used only for EG-based nanofluids.

3.5.1 Statistics

For each fluid and each temperature, we performed a large number of individual heating runs (as the ones displayed in Figure 3.2) using different heating currents that ranged between 260 and 360 mA. Typically, the number of individual heating runs was between 100 and 400. Between heating runs, a waiting time of 3 min was established for the system to return to equilibrium. From each heating run, i, the electrical resistance $R_{0,i}$ at the beginning of the heating, the current I_i, and the asymptotic slope b_i of the $\{V, \ln t\}$ curve were stored in a computer file. The large number of individual runs allow us to perform detailed statistics. As an example, we show in Figure 3.4 histograms of the distribution of $I^3 R_0/b$ values for two representative cases, for which 223 individual heating runs were performed for water (right panel) and 272 for EG (left panel). Superimposed to the histograms in Figure 3.4, we show a fit to (non-normalized) gaussians. A simple look at the figure shows that, in both cases, the distribution of experimental points is reasonably well explained by gaussian functions, which confirms the adequacy of our experimental design. In particular, it is worth mentioning that Figure 3.4 includes $I^3 R_0/b$ values obtained at six different current intensities between 260 and 360 mA. The fact that the measured $I^3 R_0/b$ values spread around a common value for each fluid, irrespective of the applied current, confirms the adequacy of data obtained via Equation 3.8, and enhances the reliability of the λ values.

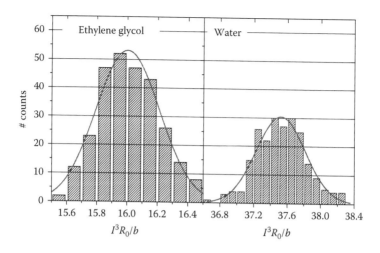

FIGURE 3.4

Histograms of experimental I^3R_0/b values obtained from individual heating runs binned at 0.1 A² intervals. Right panel is for water at 40.2°C average temperature, and left panel is for ethylene glycol at 39.3°C average temperature. Both histograms contain data obtained with the same wire at six different values of I. Superimposed are gaussian fittings.

All the thermal conductivity values quoted in the present paper were obtained by performing statistical analysis similar to the ones depicted in Figure 3.4. Hence, for each calibration or nanofluid measurement series, histograms of the distribution of experimental I^3R_0/b values obtained from individual heating runs were built and fitted to gaussians. From this fitting, we obtained an average value $\langle I^3R_0/b\rangle$, together with the corresponding variance, for each fluid and each temperature.

The water and/or EG histograms were used to infer a value of the calibration constant A by comparing with the tabulated values of the thermal conductivity of these fluids. As an example, we show in Figure 3.5 the λ values of water and ethylene glycol, as a function of temperature, obtained from the corresponding $\langle I^3R_0/b\rangle$ and adopting a single value of A. Several literature data are also shown for comparison. The λ values shown in Figure 3.5 refer to one of the wires used in our investigation, for which calibration runs were performed with both base fluids. Error bars in Figure 3.5 are the variances of the corresponding gaussian fittings (see Figure 3.4) plus a 2.5% that represents the contribution of systematic errors. Figure 3.5 demonstrates the goodness of the calibration procedure and demonstrates that our analysis method gives values for λ within the quoted accuracy.

3.5.2 Nanofluids

In the case of nanofluids, we are interested in the thermal conductivity enhancement $\Delta\lambda$, defined by Equation 3.9. In view of the definition and

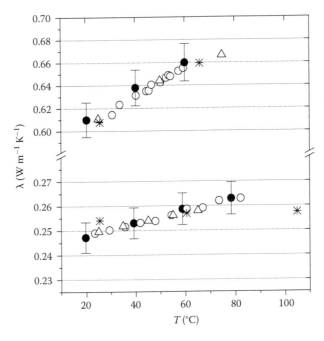

FIGURE 3.5
Comparison of the thermal conductivity values obtained in this research (filled circles) with several literature values. For water (top panel), triangles are from the ASTM-D2717 standard, empty circles are from Assael et al. [28], and asterisks from Bohne et al. [29]. For EG (bottom panel), triangles are from Khayet and Ortiz de Zárate [14], empty circles from Assael et al. [28], and asterisks from DiGuilio and Teja [30]. (Reprinted with permission from Vázquez Peñas, J.R., Ortiz de Zárate, J.M., and Khayet, M., Measurement of the thermal conductivity of nanofluids by the multicurrent hot-wire method, *J. Appl. Phys.*, 104, 044314, 2008. Copyright 2008, American Institute of Physics.)

according to Equation 3.8, $\Delta\lambda$ does not initially depend on the calibration constant A. Then, we obtained the enhancement comparing the averages $\langle I^3 R_0/b \rangle$ from the histograms of the nanofluid and the corresponding base fluid at the same bath temperature. We note that small temperature differences between the actual average temperatures measured for the pure liquid and for the corresponding nanofluids were unavoidable. In any case, these differences were less than ± 0.1 K always.

In Table 3.2, we review the data obtained for the effective thermal conductivity enhancement in our laboratory during the last years. As indicated at the header of each one of the various subtables, nanofluids based on water and nanofluids based on ethylene glycol are reported in the same table. For the evaluation of the experimental $\Delta\lambda$, we have used as reference, for each nanofluid at each temperature, the pure liquid T data obtained at the same bath temperature. Temperatures reported in Table 3.2 are the average of the

TABLE 3.2

Estimated Values of the Volume Fraction of Particles (ϕ) and Experimentally Measured Values of the Thermal Conductivity Enhancement ($\Delta\lambda$) for the Various Nanofluids Investigated in Our Laboratory

W1: SiO$_2$		W2: CuO		W3: CuO		EG1: SiO$_2$	
$\phi \simeq 2.2\%$		$\phi \simeq 0.4\%$		$\phi \simeq 0.8\%$		$\phi \simeq 1.2\%$	
T (°C)	$\Delta\lambda$	T (°C)	$\Delta\lambda$	T (°C)	$\Delta\lambda$	T (°C)	$\Delta\lambda$
20.4	3.04	20.3	0.44	20.3	1.59	19.9	0.79
44.8	3.49	40.1	1.11	40.0	2.92	39.7	0.58
60.5	1.41	60.3	2.95	60.1	4.65	59.0	1.53
—	—	—	—	—	—	79.0	1.47
Avg.	2.65	Avg.	1.50	Avg.	3.05	Avg.	1.09
EG2: SiO$_2$		**EG3: CuO**		**EG4: CuO**		**EG5: AlN**	
$\phi \simeq 2.5\%$		$\phi \simeq 0.4\%$		$\phi \simeq 0.8\%$		$\phi \simeq 0.5\%$	
T (°C)	$\Delta\lambda$	T (°C)	$\Delta\lambda$	T (°C)	$\Delta\lambda$	T (°C)	$\Delta\lambda$
20.6	3.61	20.3	2.51	20.6	5.85	19.8	2.66
40.5	4.13	39.7	3.06	42.3	6.01	39.1	2.04
60.5	4.42	58.4	4.22	60.1	6.27	58.7	2.19
80.2	4.42	80.1	3.20	79.8	5.35	78.1	2.91
Avg.	4.14	Avg.	3.25	Avg.	5.87	Avg.	2.45

Note: Thermal conductivity enhancements are reported in %. See text for comments on uncertainties.

readings performed by the reference thermometer before each individual heating run.

For the sake of clarity, we do not explicitly report errors in the data displayed in Table 3.2. The variances obtained from the histogram fitting procedure were in all cases less than 0.1%. Furthermore, values of the thermal conductivity enhancement $\Delta\lambda$ have been obtained by the calibration method so that we estimate they are mostly free from systematic errors, up to 0.5%. As a conclusion, we estimate the uncertainty of the values reported in Table 3.2 as ± 0.01 (%) for all the $\Delta\lambda$ values displayed.

3.6 Discussion

3.6.1 Comparison with Available Experimental Data

For a given nanofluid, the experimental enhancement $\Delta\lambda$, as reported in Table 3.2, is roughly independent of the temperature. For the nanofluids obtained by dispersing cooper oxide nanoparticles in water, it seems that there is a systematic increase of $\Delta\lambda$ with the temperature, but this trend is not confirmed in the nanofluids with SiO_2 or AlN nanoparticles, nor in the nano-fluids with CuO particles dispersed in ethylene glycol. In any case, the sta-tistical significance of this trend is very marginal since the apparent increase in the CuO/water nanofluid is almost within the experimental uncertainty of the measurements. Some authors [31–33] report $\Delta\lambda$ increase with tempera-ture, as marginally shown here by the CuO/water nanofluid data. However, in overall, from our present data and for the temperature range studied, our conclusion is that $\Delta\lambda$ is independent of the temperature within the accuracy of our measurements. As a consequence, in the last rows of Table 3.2, we have evaluated average enhancements for each nanofluid and we continue our discussion in terms of these average enhancements.

For the nanofluids that have been measured at more than a single vol-ume fraction ϕ of nanoparticles, there is a clear increase in $\Delta\lambda$ with ϕ, as shown in Figure 3.6 for the nanofluids containing CuO nanoparticles. This trend has been consistently observed by other investigators in the field. In a previous publication [24], we have reported more in detail on

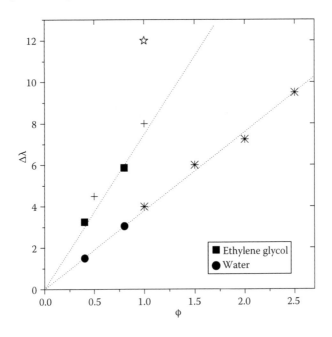

FIGURE 3.6
Comparison with literature data of the thermal conductivity enhancements measured in this research for the CuO nanofluids (filled symbols, with base fluids as indicated). Asterisks rep-resent data from Lee et al. [34] for CuO/water nanofluids. Crosses represent data from Patel et al. [36] for CuO/ethylene glycol nanofluids at 20°C. Finally, the five-pointed star repre-sents the lower volume fraction value of the pioneering data of Eastman et al. [35], also for a CuO/ethylene glycol nanofluid.

this behavior and compared with literature data published for the same systems and available at that time [6,34,35].

It is worth mentioning that our results for $\Delta\lambda$ are significantly lower than some other published data. For instance, the enhancement observed for CuO/water nanofluids is about half the one reported by Li and Peterson [32] for 19 nm CuO nanoparticles or by Patel et al. [36] for particles of 31 nm. Similar differences exist with values reported by Das et al. [31]. In these three cases, the difference may be related to the different methods followed for estimating ϕ. On the other hand, our $\Delta\lambda$ values for CuO/water nanofluids fully agree with those of Lee et al. [34] or with those of Zhu et al. [37] for the same system at similar ϕ, as shown in Figure 3.6. Regarding the CuO/EG nanofluids, our $\Delta\lambda$ values are consistent with those reported by Patel et al. [36] for the same system at similar ϕ, as well as with those of Kwak and Kim [6], or, more recently, with the ones by Longo and Zilio [33] for TiO_2/EG at similar ϕ (CuO and TiO_2 have a similar intrinsic thermal conductivity). However, our CuO/EG enhancements clearly disagree with the pioneering measurements of Eastman et al. [35]. In Figure 3.6, we summarize this comparison of our present results with the literature for nanofluids containing CuO nanoparticles. We show our present results (filled symbols) together with some literature data that agree with ours and some literature data that do not agree (usually larger than what we obtain).

With respect to the SiO_2/W nanofluid (for which we measure a single ϕ), the experimental value we have obtained is compatible with those reported by Kang and Kim [38] or, most recently, by Chen et al. [39] for similar ϕ. However, our value is about half that reported by Hwang et al. [27] while it is about twice the value measured by Tavman et al. [8]. Unfortunately, this situation is not uncommon in nanofluid research and is somewhat hampering progress in the field.

Regarding the AlN/EG nanofluid, fewer experimental data are available so far, and in refereed scientific journals, we could only find measurements by Yu et al. [9]. Our single value at $\phi = 0.5\%$ is consistent with the data reported by them, accounting for the much larger ϕ range investigated elsewhere [9] (up to $\phi = 10.0\%$).

3.6.2 Effective Medium Theoretical Models

One interesting issue concerning the thermal conductivity of nanofluids is if the observed enhancements can be explained in terms of existing effective-medium theoretical models. Recently, Wang and Mujumdar [26] have extensively reviewed different theories that have been proposed to explain the thermal conductivity of suspensions of particles in fluids. In addition to all the specific models described elsewhere [26], we think it is quite insightful to consider also the naive idea of modeling the nanofluid as a series of two layers, one of nanoparticles and the other of pure fluid. In this case, a very simple exercise shows that the effective thermal conductivity of the composite layer can be expressed as

$$\frac{1}{\lambda_{\text{eff}}} = \frac{\phi}{\lambda_p} + \frac{1-\phi}{\lambda_f} \tag{3.10}$$

where
  λ_p is the thermal conductivity of the nanoparticles
  λ_f is the thermal conductivity of the pure liquid

The first among the more sophisticated models is the adaptation by Hamilton and Crosser [11] of the classical Maxwell model [10], initially developed to predict the dielectric properties of a suspension of microspheres. This model predicts an effective thermal conductivity as [12,26]

$$\frac{1}{\lambda_{\text{eff}}} = \frac{1}{\lambda_f} - \frac{\phi}{\lambda_f} \frac{3\left(\lambda_p - \lambda_f\right)}{\lambda_p + 2\lambda_f + 2\left(\lambda_p - \lambda_f\right)\phi} \tag{3.11}$$

A generalization of the Maxwell, Hamilton, and Crosser [10,11] model has been developed by Nan et al. [40], but still being an effective-medium theory. Nan et al. [40] account for nonspherical particles and for interfacial thermal resistance. We have verified, however, that if we consider our nanoparticles spherical, the possible interfacial thermal resistance of Nan et al. [40] gives negligible differences with the Hamilton and Crosser model, predicting slightly lower enhancements. Similarly, as also reviewed by Wang and Mujumdar [26], for small concentrations in volume fraction (as the data we present here), there is little difference between the classical Hamilton and Crosser model and the other more sophisticated theories reviewed by them.

We should mention that, in addition to the analytic models reviewed by Wang and Mujumdar [26], more realistic theoretical predictions can be developed by specific computer simulations, a field that has been particularly active during last years [41,42]. Obviously, by numerical simulations, features that are hard to treat analytically can be introduced in the theory, but at the expense of elaborating very specific models that are not easily translated from one system to other.

In any case, since our purpose here is to evaluate up to what point the measured enhancements $\Delta\lambda$ can be considered anomalous, in addition to the two-layer model of Equation 3.10, we shall compare our current experimental data with the classical Hamilton and Crosser model [11] of Equation 3.11 only. In Table 3.3, we show the experimental values of the temperature-averaged effective thermal conductivity enhancement $\Delta\lambda_{\text{exp}}$ (already displayed in the last rows of Table 3.2), together with the predictions based on the two-layer model of Equation 3.10 and the model of Equation 3.11, for the various nanofluids studied in the present investigation. To compute the model's predictions, we took the pure fluid's thermal conductivity from the

TABLE 3.3

Comparison between the Experimental Enhancements, $\Delta\lambda_{exp}$, and the Predictions Based on the Two-Layer Model of Equation 3.10, $\Delta\lambda_{2lay}$, and the Hamilton and Crosser Model of Equation 3.11, $\Delta\lambda_{HC}$

Nanofluid	$\Delta\lambda_{exp}$ (%)	$\Delta\lambda_{2lay}$ (%)	$\Delta\lambda_{HC}$ (%)
SiO$_2$/W			
$\phi = 2.2\%$	2.65	1.19	1.84
CuO/W			
$\phi = 0.4\%$	1.50	0.39	1.15
$\phi = 0.8\%$	3.06	0.79	2.30
SiO$_2$/EG			
$\phi = 1.2\%$	1.09	0.98	2.13
$\phi = 2.5\%$	4.14	2.07	4.47
CuO/EG			
$\phi = 0.4\%$	3.25	0.40	1.18
$\phi = 0.8\%$	5.87	0.80	2.36
AlN/EG			
$\phi = 0.5\%$	2.45	0.50	1.50

Note: Different nanofluids are indicated. All enhancements are reported in %.

data at 20°C obtained during the calibration procedure. For the thermal conductivity of the nanoparticles, we used the tabulated values for the bulk solids: $\lambda_{SiO_2} \simeq 1.3\,\mathrm{W\,m^{-1}\,K^{-1}}$ (noncrystalline or fused silica), $\lambda_{CuO} \simeq 33\,\mathrm{W\,m^{-1}\,K^{-1}}$, and $\lambda_{AlN} \simeq 200\,\mathrm{W\,m^{-1}\,K^{-1}}$. We note that nanoparticles for our investigation cover a wide range of intrinsic thermal conductivities.

An inspection of the values reported in Table 3.3 shows that the theoretical enhancements calculated using the two-layer model of Equation 3.10 are in all cases significantly smaller than the experimental enhancements. Regarding the estimations based on the effective medium model of Hamilton and Crosser, we observe that for the nanofluids containing silica nanoparticles (SiO$_2$/W, SiO$_2$/EG), Equation 3.11 gave quite reasonable estimations, in agreement with what has been recently found for alumina particles [13]. However, for the nanofluids containing CuO or AlN nanoparticles, the estimations based on the Hamilton and Crosser model somewhat underestimated the measured enhancements. Although these results may suggest an *anomalous* $\Delta\lambda$ for the nanofluids containing particles of larger intrinsic λ, we stress that the anomaly we have found is quite small, far from the large amounts originally claimed by Eastman et al. [5]. In fact, the anomaly is in our case barely above of what can be considered as statistically significant. It is obvious that more experimental efforts are required to confirm whether the claimed anomaly is real, and if the effects of Brownian motion, cluster formation, etc. [12,26], are really relevant. We look forward to seeing more developments in this extremely interesting topic in the coming future.

Acknowledgments

We have benefitted from discussions with many colleagues during the years that took this investigation. In particular, we thank M. Piñeiro and L. Lugo, from the University of Vigo, and M. Vallés from the Rovira i Virgili University (Tarragona). The authors gratefully acknowledge the financial support of the I+D+i Project MAT2010-19249 (Spanish Ministry of Science and Innovation).

References

1. L. Wang, J. Fan, Nanofluids research: Key issues, *Nanoscale Res. Lett.* 5 (2010): 1241–1252.
2. H. Xie, W. Yu, Y. Li, L. Chen, Discussion on the thermal conductivity enhancement of nanofluids, *Nanoscale Res. Lett.* 6 (2011): 124.
3. S. W. Hong, Y.-T. Kang, C. Kleinstreuer, J. Koo, Impact analysis of natural convection on thermal conductivity measurements of nanofluids using the transient hot-wire method, *Int. J. Heat Mass Transfer* 54 (2011): 3448–3456.
4. J. Philip, P. D. Shima, Thermal properties of nanofluids, *Adv. Colloid Interface Sci.* 183–184 (2012): 30–45.
5. J. A. Eastman, S. U. S. Choi, S. Li, W. Yu, L. J. Thompson, Anomalously increased effective thermal conductivities of ethylene glycol-based nanofluids containing copper nanoparticles, *Appl. Phys. Lett.* 78 (2001): 718–720.
6. K. Kwak, C. Kim, Viscosity and thermal conductivity of copper oxide nanofluid dispersed in ethylene glycol, *Korea—Aust. Rheol. J.* 17 (2005): 35–40.
7. M. J. Pastoriza-Gallego, C. Casanova, R. Páramo, B. Barbés, J. L. Legido, M. M. Piñeiro, A study on stability and thermophysical properties (density and viscosity) of Al_2O_3 in water nanofluid, *J. Appl. Phys.* 106 (2009): 064301.
8. I. Tavman, A. Turgut, M. Chirtoc, H. P. Schuchmann, S. Tavman, Experimental investigation of viscosity and thermal conductivity of suspensions containing nanosized ceramic particles, *Arch. Mater. Sci. Eng.* 34 (2008): 99–104.
9. W. Yu, H. Xie, Y. Li, L. Chen, Experimental investigation on thermal conductivity and viscosity of aluminum nitride nanofluid, *Particuology* 9 (2011): 187–191.
10. J. C. Maxwell, *A Treatise on Electricity and Magnetism*, 2nd edn., Clarendon Press, Oxford, U.K., 1881.
11. R. L. Hamilton, O. K. Crosser, Thermal conductivity of heterogeneous 2-component systems, *Ind. Eng. Chem.: Fundamen.* 1 (1962): 187–191.

12. K. S. Gandhi, Thermal properties of nanofluids: Controversy in the making?, *Curr. Sci.* 6 (2007): 717–718.

13. J. Buongiorno, D. C. Venerus, N. Prabhat, T. McKrell, J. Townsend, R. Christianson, Y. V. Tolmachev et al., A benchmark study on the thermal conductivity of nanofluids, *J. Appl. Phys.* 106(9) (2009): 094312.

14. M. Khayet, J. M. Ortiz de Zárate, Application of the multi-current transient hot–wire technique for absolute measurements of the thermal conductivity of glycols, *Int. J. Thermophys.* 26 (2005): 637–646.

15. M. J. Assael, K. D. Antoniadis, W. A. Wakeham, Historical evolution of the transient hot-wire technique, *Int. J. Thermophys.* 31 (2010): 1051–1072.

16. Y. Nagasaka, A. Nagashima, Simultaneous measurement of the thermal conductivity and the thermal diffusivity of liquids by the transient hot-wire method, *Rev. Sci. Instrum.* 52 (1981): 229.

17. M. J. Assael, E. Charitidou, C. Nieto de Castro, W. Wakeham, The thermal conductivity of n-hexane, n-heptane and n-decane by the transient hot wire method, *Int. J. Thermophys.* 8 (1987): 663.

18. H. S. Carslaw, J. C. Jaeger, *Conduction of Heat in Solids*, Oxford University Press, Oxford, U.K., 1959.

19. M. Abramowitz, I. A. Stegun, *Handbook of Mathematical Functions*, Dover, New York, 1964.

20. J. Kestin, W. A. Wakeham, Contribution to theory of transient hot–wire technique for thermal-conductivity measurements, *Physica A* 92 (1978): 102–116.

21. C. A. Nieto de Castro, R. Perkins, H. M. Roder, Radiative heat transfer in transient hot-wire measurements of thermal conductivity, *Int. J. Thermophys.* 12 (1991): 985.

22. M. J. Assael, L. Karagiannidis, S. M. Richardson, W. A. Wakeham, Compression work using the transient hot-wire method, *Int. J. Thermophys.* 13 (1992): 223.

23. M. J. Assael, K. D. Antoniadis, K. E. Kakosimos, I. N. Metaxa, An improved application of the transient hot-wire technique for the absolute accurate measurement of the thermal conductivity of pyroceram 9606 up to 420 K, *Int. J. Thermophys.* 29 (2008): 445–456.

24. J. R. Vázquez Peñas, J. M. Ortiz de Zárate, M. Khayet, Measurement of the thermal conductivity of nanofluids by the multicurrent hot-wire method, *J. Appl. Phys.* 104 (2008): 044314.

25. J. M. Ortiz de Zárate, J. Luis Hita, M. Khayet, J. L. Legido, Measurement of the thermal conductivity of clays used in pelotherapy by the multi-current hot-wire technique, *Appl. Clay Sci.* 50 (2010): 423–426.

26. X.-Q. Wang, A. S. Mujumdar, Heat transfer characteristics of nanofluids: A review, *Int. J. Thermal Sci.* 46 (2007): 1–19.

27. Y. Hwang, J. K. Lee, C. H. Lee, Y. M. Jung, S. I. Cheong, C. G. Lee, B. Ku, S. P. Jang, Stability and thermal conductivity characteristics of nanofluids, *Thermochim. Acta* 455 (2007): 70–74.

28. M. J. Assael, E. Charitidou, S. Augustinianus, W. A. Wakeham, Absolute measurements of the thermal conductivity mixtures of alkene-glycols with water, *Int. J. Thermophys.* 10 (1989): 1127.

29. D. Bohne, S. Fischer, E. Obermeier, Thermal conductivity, density, viscosity and Prandtl numbers of ethylene glycol–water mixtures, *Ber-Bunsenges. Phys. Chem.* 88 (1984): 739.

30. R. DiGuilio, A. S. Teja, Thermal-conductivity of poly(ethylene glycols) and their binary-mixtures, *J. Chem. Eng. Data* 35 (1990): 117–121.
31. S. K. Das, N. Putra, P. Thiesen, W. Roetzel, Temperature dependence of thermal conductivity enhancement for nanofluids, *J. Heat Transfer* 125 (2003): 567–574.
32. C. H. Li, G. P. Peterson, Experimental investigation of temperature and volume fraction variations on the effective thermal conductivity of nanoparticle dispersions (nanofluids), *J. Appl. Phys.* 99 (2006): 084314.
33. G. A. Longo, C. Zilio, Experimental measurements of thermophysical properties of Al_2O_3- and TiO_2-ethylene glycol nanofluids, *Int. J. Themophys.* 34 (2013): 1288–1307.
34. S. Lee, S. U. S. Choi, S. Li, J. A. Eastman, Measuring thermal conductivity of fluids containing oxide nanoparticles, *J. Heat Transfer* 121 (1999): 280–289.
35. J. A. Eastman, S. U. S. Choi, S. Li, L. J. Thompson, S. Lee, Enhanced thermal conductivity through the development of nanofluids, in: *Materials Research Society Symposium—Proceedings*, vol. 457, Materials Research Society, Pittsburgh, PA, 1997, pp. 3–11.
36. H. E. Patel, T. Sundararajan, S. K. Das, An experimental investigation into the thermal conductivity enhancement in oxide and metallic nanofluids, *J. Nanoparticle Res.* 12 (2010): 1015–1031.
37. H. Zhu, D. Han, Z. Meng, D. Wu, C. Zhang, Preparation and thermal conductivity of CuO nanofluid via a wet chemical method, *Nanoscale Res. Lett.* 6 (2011): 181.
38. H. U. Kang, S. H. Kim, Estimation of thermal conductivity of nanofluids using an experimental effective particle volume, *Exp. Heat Transfer* 19 (2006): 181–191.
39. Y. J. Chen, P. Y. Wang, Z. H. Liu, Application of water-based SiO_2 functionalized nanofluid in a loop thermosyphon, *Int. J. Heat Mass Transfer* 56 (2013): 59–68.
40. C. W. Nan, R. Birringer, D. R. Clarke, H. Gleiter, Effective thermal conductivity of particulate composites with interfacial thermal resistance, *J. Appl. Phys.* 81 (1997): 6692–6699.
41. M. J. Assael, I. N. Metaxa, K. Kakosimos, D. Constantinou, Thermal conductivity of nanofluids—Experimental and theoretical, *Int. J. Thermophys.* 27 (2006): 999–1017.
42. J. Fan, L. Wang, Heat conduction in nanofluids: Structure–property correlation, *Int. J. Heat Mass Transfer* 54 (2011): 4349–4359.

4

Two-Phase Laminar Mixed Convection
Al₂O₃–Water Nanofluid in Elliptic Duct

Buddakkagari Vasu and Rama Subba Reddy Gorla

CONTENTS

4.1 Background and Introduction 101
    4.1.1 Background... 101
    4.1.2 Introduction... 104
4.2 Mathematical Model.. 106
    4.2.1 Mixture Model 106
    4.2.2 Properties of Nanofluid 108
    4.2.3 Boundary Conditions... 110
    4.2.4 Physical Quantities 111
    4.2.5 Special Cases 111
4.3 Numerical Simulation with Ansys Fluent 112
    4.3.1 Computational Details.. 112
    4.3.2 Validation... 112
4.4 Concluding Remarks.. 113
4.5 Research Opportunities... 115
Nomenclature 116
References.. 117

4.1 Background and Introduction

4.1.1 Background

The heat convection can passively be enhanced by changing flow geometry and boundary conditions or by enhancing fluid thermophysical properties. One way is by adding small solid particles in the fluid. The main idea backs Maxwell's study [1]. He presented the possibility of growing thermal conductivity of a fluid–solid mixture by more volume fraction of solid particles. The particles with micrometer or even millimeter dimensions were used. Those particles caused several problems such as abrasion, clogging, and pressure dropping. By developing the technology to make particles in nanometer dimensions, a new generation of solid–liquid mixtures called nanofluid was created. Nanofluid is a suspension of nanoparticles such as Al_2O_3, Cu, or CuO in a base fluid such as water, ethylene glycol, or oil. Nanofluids are a new kind of heat transfer fluids containing a small quantity of nanosized particles (usually less than 100 nm) that are uniformly and stably suspended in a liquid. The dispersion of a small amount of solid nanoparticles in conventional fluids changes their thermal conductivity remarkably. Anomalous thermal conductivity enhancement in nanofluid suspension was reported by Choi et al. [2]. Xuan and Roetzel [3] have identified two causes of improved heat transfer by nanofluids: the increased thermal dispersion due to the chaotic movement of nanoparticles that accelerates energy exchanges in the fluid and the enhanced thermal conductivity of nanofluids. The use of nanofluids is one of the most effective mechanisms of increasing the amount of heat transfer in heat exchangers.

Multiphase flow played a fundamental role for the duration of this study since the concepts involved in this topic dictated many of the preliminary calculations and assumptions made from which to base results. Multiphase flow is a flow with simultaneous presence of different phases, where phase refers to solid, liquid, or vapor state of matter. There are four main categories of multiphase flows: gas–liquid, gas–solid, liquid–solid, and three-phase flows. Further characterization is commonly done according to the visual appearance of the flow as separated, mixed, or dispersed flow. These are called flow patterns or flow regimes, and the categorization of a multiphase flow in a certain flow regime is comparable to the importance of knowing if a flow is laminar or turbulent in single-phase flow analysis (Thome [4]).

A flow pattern describes the geometrical distribution of the phases and greatly affects phase distribution and velocity distribution for a certain flow situation. A number of flow regimes exist, and the possible flow patterns differ depending on the geometry of the flow domain. For some simple shapes, for example, horizontal and vertical pipes, the flow patterns that occur for different phase velocities, etc., have been summarized in a so-called flow map. Figure 4.1 visualizes the flow configuration for some possible flow regimes, and Figure 4.2 shows an example of a flow map for horizontal pipe flow.

The two extremes on a flowmap are dispersed flow and separated flow. In separated flow, there is a distinct boundary between the phases. Several intermediate regimes also exist, which contain both separated and dispersed phases such as annular bubbly flow. Due to growing instabilities in one regime, transition to another regime can occur. This phenomenon complicates the modeling of multiphase flow even further as the transition is unpredictable and the different flow regimes are to some extent governed by different physics. Multiphase flow is applicable in a variety of operations, including reaction processes, separations, and purification processes.

Convective heat transfer with nanofluids can be modeled using the two-phase or single-phase approach. The first provides the possibility of

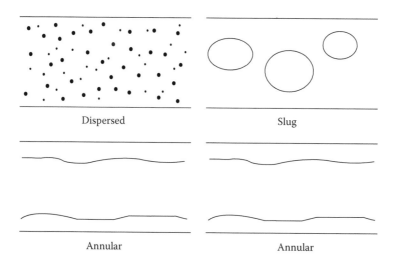

FIGURE 4.1
Example of typical flow patterns for flow in horizontal pipes.

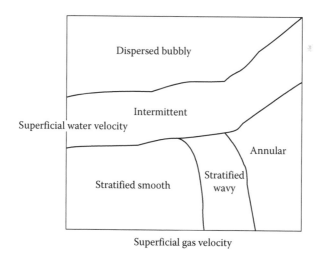

FIGURE 4.2
Example of flow map for two-phase flow in horizontal pipes.

understanding the functions of both the fluid phase and the solid particles in the heat transfer process. The second assumes that the fluid phase and particles are in thermal equilibrium and move with the same velocity. This approach is simpler and requires less computational time. Thus, it has been used in several theoretical studies of convective heat transfer with nanoflu-ids. Mixed convection in horizontal tubes at macro size has been of interest in many industries. Therefore, it has been studied extensively. Heat transfer is enhanced in such a condition because of secondary flow that is generated by the buoyancy force. Buoyancy force significantly affects the flow field; for instance, at the fully developed region, the maximum axial velocity does not appear at the tube center line, and radial variation of the temperature becomes important.

4.1.2 Introduction

Nanofluid transport phenomena have received extensive attention in the recent years. Although interests in nanofluids were initially directed at thermal enhancement in transport engineering (automotive, aerospace, etc.), applications have significantly diversified in the past decade. New areas in which nanofluids have been deployed include electronics cooling [5], building physics and contamination control [6], pharmacological administration mechanisms [7], peristaltic pumps for diabetic treatments [8], solar collectors [9], and bioconvection in microbial fuel cells [10].

A comprehensive survey of convective transport in nanofluids has been made by Buongiorno [11] who identified *Brownian diffusion* and *thermophoresis* as key mechanisms contributing to thermal conductivity enhancement. Boundary layer flows in porous media were first investigated by Nield and Kuznetsov [12] using similarity transformations. Rashidi et al. [13] used the homotopy analysis method (HAM) and Mathematica symbolic software to analyze coating flows of a vertical cylinder with nanofluids. Gorla et al. [14] have studied mixed convective boundary layer flow over a vertical wedge embedded in a porous medium saturated with a nanofluid: natural convection dominated regime. These studies all highlighted the significant role of thermophoresis and Brownian motion in enhancing heat transfer rates. Experimental works have been complemented in recent years by numerous theoretical and computational simulations. The latter have deployed an extensive range of sophisticated algorithms.

Keblinski et al. [15] have studied four possible mechanisms that contribute to the increase in nanofluid heat transfer: Brownian motion of the nanoparticles, molecular-level layering of the base fluid/nanoparticle interface, heat transport in the nanoparticles, and nanoparticle clustering. Due to the many advantages of using nanofluids, researchers have conducted numerous experimental and numerical studies on nanofluids in recent years [16–18]. In addition to costly experimental studies, the numerical simulation of nanofluids using computational fluid dynamics (CFD) techniques is another effective approach in analyzing the performance of nanofluids [19,20]. In general, for the numerical simulation of nanofluids, there are two methods called the single-phase and two-phase, with the two-phase method being much more exact [21]. The single-phase approach has been used in several theoretical studies of convective heat transfer with nanofluids [22–29]. Because the properties of nanofluids are not completely specified, and there are not good expressions for predicting nanofluid mixture, the single-phase numerical predictions are not generally in good agreement with experimental results.

On the other hand, because the two-phase approach considers the movement between the solid and fluid molecules, it is a more accurate approach to nanofluid study. To fully describe and predict the flow and behavior of complex flows, different multiphase theories have been proposed and used. The two-phase method itself has different categories, including the Eulerian–Eulerian method, mixture method, etc. Using the three methods of two-phase Eulerian–Eulerian, two-phase mixture, and single-phase homogeneous, Lotfi et al. [30] simulated the Al_2O_3–water nanofluid flow in circular tubes. The large number of published articles concerning multiphase flows typically employed the mixture theory or the theory of interacting continua [31,32]. This approach is based on the underlying assumption that each phase can be mathematically described as a continuum. Some researchers have employed two-phase mixture theory to predict the behavior of nanofluids. Akbarinia and Laur [33] and Mirmasoumi and Behzadmehr [34] studied the effect of the nanoparticles' diameter on nanofluid fluid in horizontal curved tubes with circular cross section. Mokhtari Moghari et al. [35] studied two-phase mixed convection Al_2O_3–water nanofluid flow in an annulus. By comparing the simulation results with the experimental data, they came to the conclusion that the two-phase mixture method is the most exact method among the existing approaches. Also in this article, the two-phase mixture method is used for the numerical simulation of nanofluid flow in elliptic ducts.

Elliptic cross section tubes have drawn special attention since they were found to create less resistance to the cooling fluid, which results in less pumping power [36]. Velusamy et al. [37] have studied mixed and forced convection fluid flow in ducts with elliptic and circular cross sections. They found that irrespective of the value of the Rayleigh number, the ratio of friction factor during mixed convection to corresponding value during forced convection is low in elliptic ducts compared to that in a circular duct. Also, the ratio of the Nusselt number to friction factor is higher for elliptic ducts compared to that for a circular duct. Despite the fact that the secondary flow in elliptic ducts is very small compared with the streamwise bulk flow, secondary motions play a significant role by cross-stream transferring momentum, heat, and mass. The main advantage of using elliptic ducts rather than circular ducts is the enhancing of the heat transfer coefficient [38]. The elliptic tubes are employed in many practical fields in the area of energy conservation, design of solar collectors, heat exchangers, nuclear engineering, cooling of electrical and electronic equipment, refrigeration and air-conditioning applications, and many others. Hence, heat transfer enhancement in these devices is essential, and the use of nanofluids can be effective in increasing the heat transfer coefficient.

Nanofluids have numerous industrial applications, where efficient heat dissipation is necessary. Therefore, special attention has been given to nanofluid flow in tubes with elliptic cross sections employing the two-phase mixture model. The objective of the present chapter is to study the impact of nanoparticle volume fraction, aspect ratio, and buoyancy forces on the laminar mixed convection of nanofluids flow in elliptic pipes. The proposed work will entail mathematical formulation and analyses and numerical algorithms for the computations and present and discuss the axial velocity, secondary flow pattern, contours of temperature, distribution of nanoparticles, skin friction factor, and Nusselt number profiles.

4.2 Mathematical Model

4.2.1 Mixture Model

Mixed convection of a nanofluid consisting of water and Al_2O_3 nanoparticles in horizontal elliptic ducts with uniform heat flux at the solid–liquid interface has been considered. Figure 4.3 shows the geometry of the considered problem. The computation domain is composed of a straight elliptic pipe with a length of L, a horizontal ellipse semi-axis of a, and a vertical ellipse semi-axis of b, whereas the gravitational force is exerted in the vertical direction. The nanofluid flow is laminar, steady state. The properties of the fluid are assumed constant except for the density in the body force, which varies linearly with the temperature (Boussinesq's hypothesis). The mixture model, based on a single-fluid two-phase approach, is employed in the simulation by assuming that the coupling between phases is strong, and particles closely follow the flow. The two phases are assumed to be interpenetrating, meaning that each phase has its own velocity vector field, and within any control volume, there is a volume fraction of the primary phase and also a volume fraction of the secondary phase. Instead of utilizing the governing equations of each phase separately, the continuity, momentum, and fluid

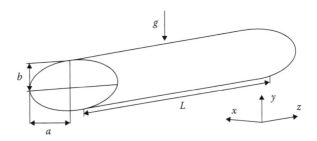

FIGURE 4.3
Problem geometry and coordinate system.

energy equations for the mixture are employed. Therefore, the steady-state governing equations describing a mixture fluid flow and the heat transfer in elliptic pipes are as follows:

Continuity equation:

$$\nabla \cdot \left(\rho_m V_m \right) = 0 \tag{4.1}$$

Momentum equation:

$$\nabla \cdot \left(\rho_m V_m V_m \right) = -\nabla_p + \nabla \cdot \left(\mu_m \nabla V_m \right) + \nabla \cdot \left(\sum_{k=1}^{n} \varnothing_k \rho_k V_{dr,k} V_{dr,k} \right) - \rho_{m,i} \beta_m g \left(T - T_i \right) \tag{4.2}$$

Fluid energy equation:

$$\nabla \cdot \sum_{k=1}^{n} \left(\rho_k C_{pk} \varnothing_k V_k T \right) = \nabla \cdot \left(k_m \nabla T \right) \tag{4.3}$$

Volume fraction equation:

$$\nabla \cdot \left(\phi_p \rho_p V_m \right) = -\nabla \cdot \left(\phi_p \rho_p V_{dr,p} \right) \tag{4.4}$$

V_m is mass average velocity,

$$V_m = \frac{\sum_{k=1}^{n} \left[\left(\rho_k \phi_k V_k \right) \right]}{\rho_m} \tag{4.5}$$

In Equation 4.2, $V_{dr,k}$ is the drift velocity for the secondary phase k, that is, the nanoparticles in the present study,

$$V_{dr,k} = V_k - V_m \tag{4.6}$$

The slip velocity (relative velocity) is defined as the velocity of a secondary phase (nanoparticles, p) relative to the velocity of the primary phase (water, f)

$$V_{pf} = V_p - V_f \tag{4.7}$$

The drift velocity is related to the relative velocity,

$$V_{dr,p} = V_{pf} - \sum_{k=1}^{n} \frac{\phi_k \rho_k}{\rho_m} V_{fk} \tag{4.8}$$

The relative velocity is determined from Equation 4.9 proposed by Manninen et al. [32] while Equation 4.10 by Schiller and Naumann [39] is used to calculate the drag function f_{drag}.

$$V_{pf} = \frac{\rho_p d_p^2}{18\mu_f f_{drag}} \frac{(\rho_m - \rho_p)}{\rho_p} a \qquad (4.9)$$

$$f_{drag} = \begin{cases} 1+0.15Re_p^0 \cdot 687 & Re_p \leq 1000 \\ 0.0183Re_p & Re_p > 1000 \end{cases} \qquad (4.10)$$

The acceleration (a) in Equation 4.9 is

$$a = g - (V_m \cdot \nabla) V_m \qquad (4.11)$$

where g is gravitational acceleration, which is applied in y direction as shown in Figure 4.3. It should be mentioned that other correlations are also found in the literature for calculating the drag coefficient, for instance [40–42]. Ossen [40] developed a correlation that is adequate for $Re_p < 0.5$. Proudman and Pearson [41] presented a higher order approximate solution, which is reasonable up to $Re_p = 4$. A more accurate expression is given by Clift and Gauvin [42] for $Re_p < 2 \times 10^5$, while the Schiller and Naumann [39] drag expression is quite simple and accurate for $Re_p < 800$. These correlations for calculating drag force were not developed for the nanosized particles. However, since Re_p is very low and the flow regime is considered laminar, we assume it is reasonable to use such a correlation in the absence of a particular correlation for the nanosized particles.

4.2.2 Properties of Nanofluid

The mixture model behaves similar to a single-phase liquid. Therefore, best expressions from the literature have been selected in order to better thermal and hydraulics prediction for Al_2O_3–water nanofluid application in an elliptic duct. The Al_2O_3 nanoparticle and water properties used for calculating the mixture properties are given in Table 4.1. The Brownian motions of

TABLE 4.1

Thermophysical Properties of Nanoparticles and Base Fluid at 20°C

Properties	Water	Nanoparticle (Al$_2$O$_3$)
Density ρ (kg/m^3)	998.2	3720
Heat capacitance C_p (J/kg·K)	4182	880
Thermal conductivity k (W/m·K)	0.6028	35

nanoparticles have been considered to determine the thermal conductivity and dynamics viscosity of Al$_2$O$_3$–water nanofluid, which take into account the dependence of these properties on temperature and nanoparticle volume fraction. The physical properties of nanoparticles with a spherical shape in Al$_2$O$_3$–water nanofluid are calculated as follows.

Effective mixture density:

$$\rho_m = (1 - \phi)\rho_f + \phi\rho_p \tag{4.12}$$

Recently, Masoumi et al. [43] have proposed an expression for predicting Al$_2$O$_3$–water nanofluid's dynamics viscosity, which is a function of temperature, mean nanoparticle diameter, nanoparticle volume fraction, nanoparticle density, and the based fluid physical properties. The expression used for calculating nanofluid viscosity is given as

$$\mu_m = \mu_f + \frac{\rho_p V_B d_p^2}{72C\delta} \tag{4.13}$$

where V_B and δ are, respectively, the Brownian velocity of nanoparticles and the distance between particles, which can be obtained from

$$V_B = \frac{1}{d_p}\sqrt{\frac{18k_B T}{\pi \rho_p d_p}} \tag{4.14}$$

$$\delta = \sqrt[3]{\frac{\pi}{6\phi}} d_p \tag{4.15}$$

C in Equation 4.13 is defined as

$$C = \mu_f^{-1}\left[(c_1 d_p + c_2)\phi + (c_3 d_p + c_4)\right] \tag{4.16}$$

where c_1, c_2, c_3, and c_4 are given as

$$c_1 = -0.000001133$$

$$c_2 = -0.000002771 \tag{4.17}$$

$$c_3 = 0.00000009$$

$$c_4 = -0.000000393$$

The thermal conductivity of Al_2O_3–water nanofluid has been determined from Chon et al. [44] correlation, which considers the Brownian motion and mean diameter of the nanoparticles as follows:

$$\frac{k_m}{k_f} = 1 + 64.7\phi^{0.7460}\left(\frac{d_f}{d_p}\right)^{0.3690} \times \left(\frac{k_p}{k_f}\right)^{0.7476} Pr_f^{0.9955} Re_f^{1.2321} \qquad (4.18)$$

where Pr_f and Re_f in Equation 4.18 are defined as

$$Pr_f = \frac{\eta}{\rho_f \alpha_f} \qquad (4.19)$$

$$Re_f = \frac{\rho k_B T}{3\pi\eta^2\lambda_f} \qquad (4.20)$$

where
  λ_f is the mean free path of water molecular ($\lambda_f = 017$ nm)
  k_B is Boltzmann constant ($k_B = 1.3807 \times 10^{-33}$ J/K)
  η has been calculated by the following equation

$$\eta = A \times 10^{B/(T-C)}, \quad A = 2.414 \times 10^{-5}, \quad B = 247.8, \quad C = 140 \qquad (4.21)$$

Thermal expansion coefficient can be calculated from the expression presented by Khanafer et al. [24] as follows:

$$\beta_m = \left[\frac{1}{1+\left\{\frac{(1-\varnothing)\rho_f}{\varnothing\rho_p}\right\}}\left(\frac{\beta_p}{\beta_f}\right) + \frac{1}{1+\left\{\frac{\varnothing}{1-\varnothing}\right\}\left(\frac{\rho_p}{\rho_f}\right)}\right]\beta_f \qquad (4.22)$$

4.2.3 Boundary Conditions

The constant heat flux boundary condition is more appropriate at the solid–fluid interface in many practical applications than the constant wall temperature, especially with laminar steady-state fluid flow in heat exchangers. Therefore, the constant heat flux boundary condition has been chosen for this study. This set of coupled nonlinear elliptic governing equations has been solved subject to the following boundary conditions:

- At the elliptic tube inlet ($z = 0$):

$$V_{m,z} = V_i, \quad V_{m,x} = V_{m,y} = 0, \quad \text{and} \quad T_m = T_i \qquad (4.23a)$$

- At the fluid wall interfaces:

$$V_{m,x} = V_{m,y} = V_{m,z} = 0; \quad q'' = k_m \left. \frac{\partial T}{\partial n} \right|_{wall} \tag{4.23b}$$

- At the elliptic tube outlet ($z = L$): $p = p_0$ and an overall mass balance correction is applied.

4.2.4 Physical Quantities

The engineering design quantities of physical interest include the Fanning-friction coefficient and local Nusselt number, which are given by

$$f = \frac{\tau_w}{(1/2)\rho_m u_i^2}, \quad \text{where } \tau_w = \mu_m \left. \frac{\partial u}{\partial y} \right|_w \tag{4.24}$$

Therefore, the local Poiseuille number (fRe) can be obtained by

$$f\,Re = 2 \frac{\partial u / \partial y|_w}{u_i} D_h \tag{4.25}$$

The local Nusselt number (Nu) can be obtained by

$$Nu = h \frac{D_h}{k}, \quad \text{where } h = \frac{q''}{T_w - T_b} \tag{4.26}$$

Therefore, the local Nusselt number for the case of uniform heat flux can be calculated from

$$Nu = \frac{q'' D_h}{k \left(T_w - T_b \right)} \tag{4.27}$$

4.2.5 Special Cases

In order to provide insight into the particular cases in the present study, we shall discuss some special cases that pertain to earlier studies in the literature.

Case I: Mirmasoumi and Behzadmehr [34] Numerical study of laminar mixed convection of a nanofluid in a horizontal tube using two-phase mixture model.

Mirmasoumi and Behzadmehr have considered the case to investigate hydrodynamic and thermal behaviors of the nanofluid consisting of water and Al₂O₃ in a horizontal tube with two-phase mixture model. Mirmasoumi and Behzadmehr cases are easily retrieved from our general model, by setting *aspect ratio* equal to 1 ($AR = 1$).

Case II: The two-phase model becomes single-phase flow when the volume fraction solid nanoparticles are given zero ($\phi = 0$).

4.3 Numerical Simulation with Ansys Fluent

4.3.1 Computational Details

The sets of coupled nonlinear differential equations are subject to the boundary equations discretized and solved numerically by Shariat et al. [45] using the control volume technique.

A second-order upwind method was used for convective and diffusive terms while the SIMPLEC procedure was introduced to couple the velocity–pressure as described by Patankar [46]. An unstructured nonuniform grid distribution has been used to discretize the computational domain as shown in Figure 4.4. It is finer near the tube entrance and near the wall where the velocity and temperature gradients are high. Several different grid distributions in cross section and in length direction have been tested to ensure that the calculated results are grid independent. It is shown in Figure 4.5 that increasing the grid numbers in any cross section and in z direction does not change significantly the dimensionless velocity and temperature. Therefore, the grid consisting of 56 in the cross section and 60 nodes in the axial direction has been selected for the present calculations.

4.3.2 Validation

In order to demonstrate the validity and also precision of the model assumptions and the numerical analysis, fully developed values of the Poiseuille numbers are compared with available numerical solutions for different aspect ratios from 0.25 to 1 by Shariat et al. [45]. The calculated Poiseuille numbers in Table 4.2 show a significant agreement with numerical results by Schenk and Han [47] and Sakalis et al. [36] at different aspect ratios. Due to comparison with single-phase flow, the volume fraction solid nanoparticles are given zero ($\phi = 0$) for all comparisons (Figure 4.5).

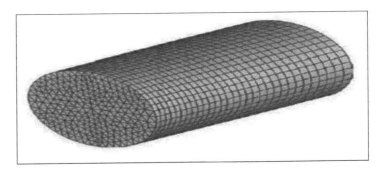

FIGURE 4.4
Unstructured nonuniform used grid for an elliptic duct.

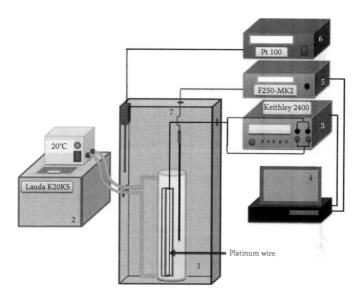

FIGURE 3.1
Schematic representation of the experimental setup for the measurement of the thermal conductivity. (1) Measurement cell, (2) Lauda K20KS thermostatic circulation bath, (3) Keithley
2400 source meter, (4) computer, (5) ASL F250 reference thermometer, (6) Pt 100 digital thermometer for monitoring chamber temperature, and (7) atmospheric chamber with controlled
temperature and humidity (Mytron).

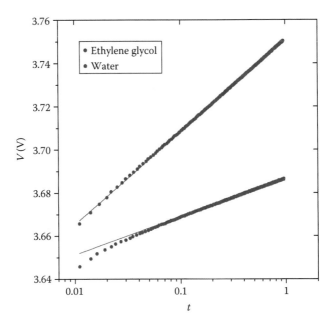

FIGURE 3.2
Typical individual heating curves for one of the platinum wires used in this research: voltage drop in the wire, V, as a function of time, t. Blue symbols are for the wire immersed in pure ethylene glycol, and red symbols for the wire immersed in pure water. Data are for an initial temperature of 40°C and heating current $I=260$ mA. Straight lines represent fittings to Equation 3.6 of the points measured after ≈213 ms.

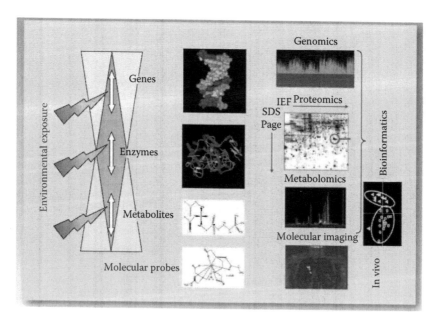

FIGURE 5.1
Molecular imaging and omics technologies. (From Spratlin, J.L. et al., *Clin. Cancer Res.*, 15(2), 431, 2009.)

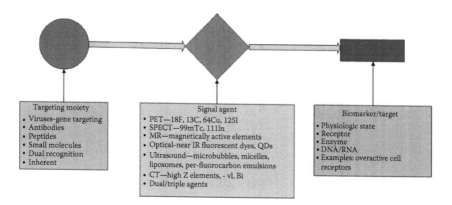

Targeting moiety
- Viruses-gene targeting
- Antibodies
- Peptides
- Small molecules
- Dual recognition
- Inherent

Signal agent
- PET—18F, 13C, 64Cu, 125I
- SPECT—99mTc, 111In
- MR—magnetically active elements
- Optical-near IR fluorescent dyes, QDs
- Ultrasound—microbubbles, micelles, liposomes, per-fluorocarbon emulsions
- CT—high Z elements, - vI, Bi
- Dual/triple agents

Biomarker/target
- Physiologic state
- Receptor
- Enzyme
- DNA/RNA
- Examples: overactive cell receptors

FIGURE 5.2
The impact of imaging in the evolutionary process of cancer.

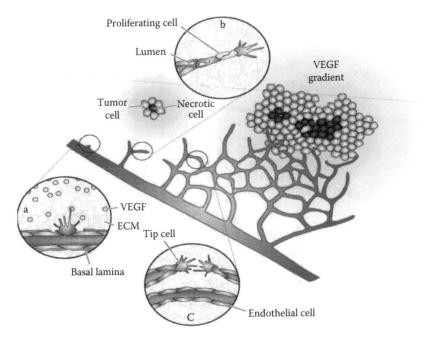

FIGURE 5.3
Steps involved in promoting angiogenesis in tumors. (From Koumoutsakos, P. et al., *Annu. Rev. Fluid Mech.*, 45, 325, 2013.)

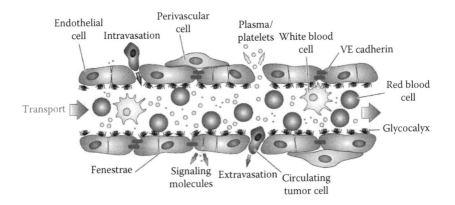

FIGURE 5.4
Tumor blood vessel structure, intravasation, transport, and extravasation of the blood constituents. (From Koumoutsakos, P. et al., *Annu. Rev. Fluid Mech.*, 45, 325, 2013.)

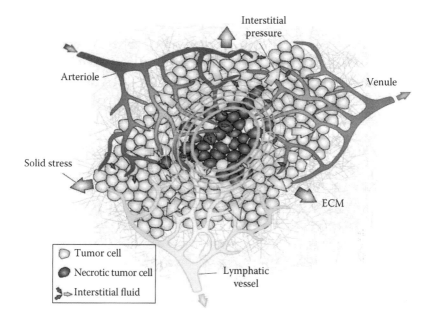

FIGURE 5.5
Landscape of the tumor microenvironment. (From Koumoutsakos, P. et al., *Annu. Rev. Fluid Mech.*, 45, 325, 2013.)

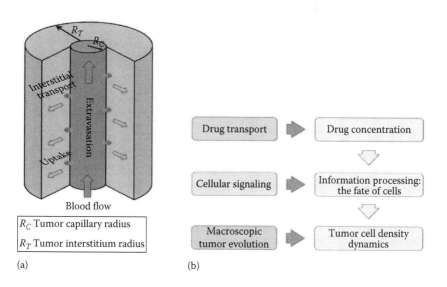

FIGURE 5.6
Information flow of the modeling framework. (a) Simplified diagram of the tumor vascular geometry and (b) flow chart of the algorithm included in the model. (From Koumoutsakos, P. et al., *Annu. Rev. Fluid Mech.*, 45, 325, 2013.)

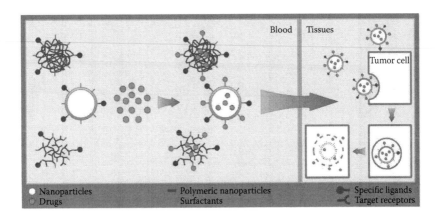

FIGURE 5.8
A schematic illustration of nanotechnology-based drug delivery systems. (From Thakor, A.S. and Gambhir, S.S., *CA Cancer J. Clin.*, 63(6), 395, 2013.)

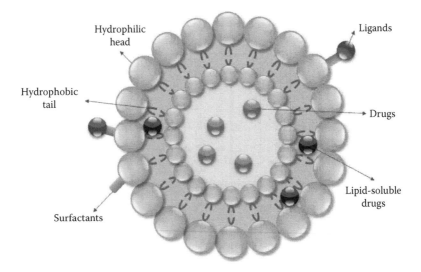

FIGURE 5.9
A schematic illustration of a liposomal formulation.

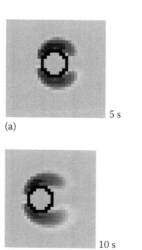

(a)   5 s

(b)   10 s

FIGURE 6.3
Simulation results from fluid dynamics simulation. (a) 5 s and (b) 10 s.

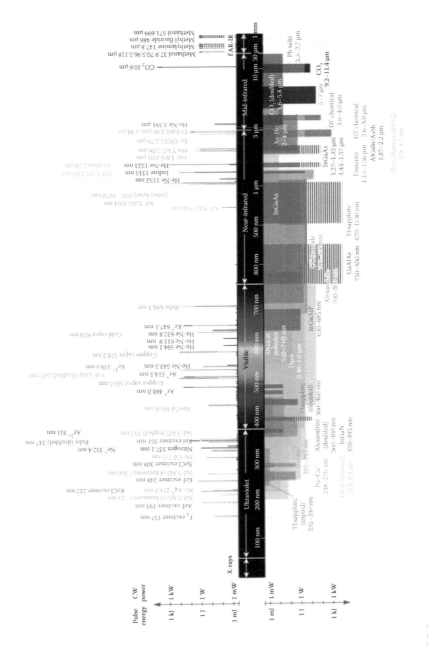

FIGURE B.2
Wavelengths of commercially available lasers. (From Weber, M.J., *Handbook of Laser Wavelengths*, CRC Press, Boca Raton, FL, 1999.)

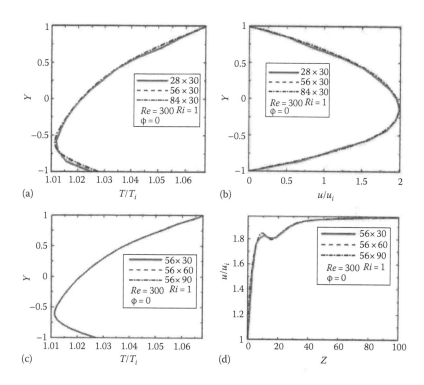

FIGURE 4.5
Grid independent test (a) and (b) in any cross section, (c) and (d) in axial direction.

TABLE 4.2

Comparison of the Poiseuille Number at Fully Developed Region with Previous Numerical Results

	$(fRe)_{fd}$		
$AR = b/a$	Numerical Result by Schenk and Han [47]	Numerical Result by Sakalis et al. [36]	Numerical Result by Shariat et al. [45]
0.25	18.29	18.258	18.36
0.5	16.823	16.896	16.85
0.75	16.317	16.255	16.25
1	16.00	16.02	16.03

4.4 Concluding Remarks

In this chapter, laminar mixed convection of nanofluids consisting of Al₂O₃–water inside tube with elliptic cross section with uniform heat flux has been presented. The effects volume fraction of nanoparticles in two constant *Re* number and constant mass flow rate was studied. The two-phase mixture model has been employed to study the solid nanoparticles (Al_2O_3) behaviors in the base fluid (water). The Brownian motions of nanoparticles have been considered to determine the thermal conductivity and dynamic viscosity of Al_2O_3–water nanofluid, which depend on the temperature.

A number of important conclusions regarding simulations of two-phase flow could be drawn from this chapter.

1. The secondary flows become strong with increasing the nanoparticles concentration, which augments the heat transfer coefficient. An increase in the nanoparticles volume fraction reduces the nondimensional temperature at a given Reynolds and Richardson number (Figures 4.6 and 4.7). At a given *Re*, increasing the nanoparticles concentration increases the Nusselt number while it causes a small decrease in the friction coefficient.

2. At constant *Re* number and heat flux with increasing nanoparticles volume fraction, the local wall temperature decreases, but the pressure drop increases. Moreover, high volume fractions increase pressure drop significantly. Also, wall temperature decreases with higher loading of nanoparticles concentration.

3. Increasing the aspect ratio of elliptic tubes shifts the velocity profile at the vertical semi-axis to the bottom wall of tubes. An increase in the aspect ratio reduces the skin friction factor remarkably. The Nusselt number does not have any specific behavior with increasing the aspect ratio, but it has maximum and minimum values for the cases of $AR = 0.75$ and $AR = 0.25$, respectively.

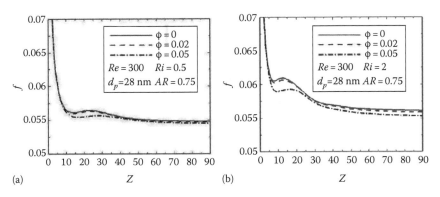

(a)

(b)

FIGURE 4.6
Variation of the friction factor with f at $Re = 300$, $AR = 0.75$, and $d_p = 28$ nm for (a) $Ri = 0.5$ and (b) $Ri = 2$.

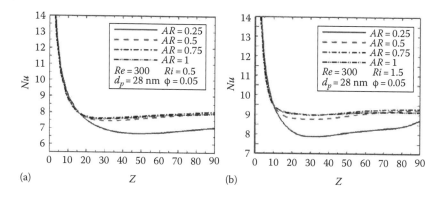

FIGURE 4.7
Variation of the local Nusselt number with solid nanoparticles volume fraction at $Re=300$, $AR=0.75$ and $d_p=28$ nm for (a) $Ri=0.5$ and (b) $Ri=1.5$.

The elliptic tube with $AR=0.75$ has almost the maximum Nusselt number and the minimum friction coefficient. Therefore, the usage of the elliptic pipes with $AR=0.75$ instead of the circular pipe is recommended.

4. As the nanoparticle volume fraction increases, pressure drop and pumping power increase.

5. By using the mixture model for simulation of nanofluid, it is observed that, though thermal properties improve, pressure drop increases, which is not desirable.

4.5 Research Opportunities

Multiphase flow played a fundamental role for the extent of this study since the concepts involved on this topic dictated many of the preliminary calculations and assumptions were made from which to base results.

In the present chapter, thermal stresses in elliptic duct at different lengths and diameters are considered due to laminar flow situations. Hence, this can be extended to turbulent and pulsating flow situations. The experimental study validating the present predictions will be fruitful for further study. Moreover, the effect of swirling at elliptic pipe inlet is not taken into account in the analysis. In some of the pipe flow applications, inlet swirling occurs because of the change in geometric configurations of the piping system. This situation can develop in the cases of flow inletting from the large-diameter elliptic pipes, inletting from large sections (flat plate heat exchangers tubing), and reservoir exiting. Consequently, when modeling the flow and temperature fields, the inlet swirling needs to be introduced in the analysis. This, in turn, results in the modification of flow and heat transfer equations employed in the simulations.

Nomenclature

a	Horizontal ellipse semi-axis (m)	Nu	Local Nusselt number $\left(=\dfrac{q''D_n}{k_m(T_w-T_b)}\right)$
a	Acceleration (m/s²)	ΔP	Pressure drop (Pa)
AR	Aspect ratio$=b/a$	P	Pressure (Pa)
b	Vertical ellipse semi-axis (m)	PP	Pumping power $\left(=\dfrac{m\Delta p_t}{\rho}\right)$ mW
C_p	Specific heat (J/kg·K)	Pr	Prandtl number $\left(=\dfrac{\alpha_m}{v_m}\right)$
D_h	Hydraulics diameter of tube (m)	q'	Uniform heat flux (W/m²)
d_f	Diameter of nanoparticles (m)	Re	Reynolds number $\left(=\dfrac{\rho_m u_i D_h}{\mu_m}\right)$
d_p	Diameter of nanoparticles (m)	Ri	Richardson number $\left(=\dfrac{G_r}{Re^2}\right)$
f	Fanning fraction coefficients $\left(=\dfrac{\tau_w}{(1/2)\rho_m u_i^2}\right)$	T	Temperature (K)
g	Gravitational acceleration (m/s²)	V	Velocity (m/s)
Gr	Grashof number $\left(=\dfrac{g\beta_m q''D_h^4}{k_m v_m^2}\right)$	x, y, z	Coordinates
k	Thermal conductivity (W/m K)	Y	Dimensionless vertical diameter $\left(=\dfrac{y}{b}\right)$
k_B	Boltzmann constant $(=1.3807\times10^{-23})$ J/K	Z	Nondimensional length $\left(\dfrac{z}{D_h}\right)$
L	Length of duct (m)		

Greek Letter

α	Thermal diffusivity $\left(= \dfrac{\mu_m}{\rho_m} \right)$		λ_f	Mean free path of water molecular (m)
β	Volumetric expansion coefficient (K^{-1})		φ	Nanoparticles volume fraction
δ	Distance between particles (m), defined in Equation 4.15		μ	Dynamic viscosity (N·s/m)
η	Variable, defined in Equation 4.21		ν	Kinematics viscosity (m^{-2}·s^{-2})
θ	Dimensionless temperature $\left(= \dfrac{T - T_i}{\left(q''D_h \right)/k_m} \right)$		ρ	Density (kg/m^3)

Subscripts

b	Bulk	*i*	Inlet condition
dr	Drift	*m*	Mixture
f	Base fluid	*p*	Particle and solid phase
k	Indices	*w*	Wall

References

1. J.C. Maxwell, *Electricity and Magnetism*, Clarendon Press, Oxford, U.K., 1873.
2. S.U.S. Choi, Z.G. Zhang. W. Yu, F.E. Lockwood, E.A. Grulke, Anomalous thermal conductivity enhancement in nano-tube suspensions, *Appl Phys Lett* 79 (2001): 2252–2254.
3. Y.M. Xuan, W. Roetzel, Conceptions for heat transfer correlation of nanofluids, *Int J Heat Mass Transfer* 43 (2000): 3701–3707.
4. J.R. Thome, *Engineering Data Book III*, Wolverine Tube, Inc., Decatur, AL, 2004.
5. J. Townsend, R.J. Christianson, Nanofluid properties and their effects on convective heat transfer in an electronics cooling application, *ASME J Therm Sci Eng Appl* 1 (2009): 031006.
6. D.P. Kulkarni, D.K. Das, R.S. Vajjha, Application of nanofluids in heating buildings and reducing pollution, *Appl Energy* 12 (2009): 2566–2573.
7. V.V. Mody, A. Cox, S. Shah, A. Singh, W. Bevins, H. Parihar, Magnetic nanoparticle drug delivery systems for targeting tumours, *Appl Nanosci* 4 (2014): 385–392.
8. O.A. Beg, D. Tripathi, Mathematica simulation of peristaltic pumping in double-diffusive convection in nanofluids: A nano bio-engineering model, *Proc IMechE Part N: J Nanoeng Nanosyst* 225 (2012): 99–114.

9. P. Rana, R. Bhargava, O.A. Beg, Finite element modeling of conjugate mixed convection flow of Al$_2$O$_3$–water nanofluid from an inclined slender hollow cylinder, *Phys Scr Proc Royal Swedish Acad Sci* 87 (2013): 1–16.

10. O.A. Beg, V.R. Prasad, B. Vasu, Numerical study of mixed bioconvection in porous media saturated with nanofluid containing oxytactic micro-organisms, *J Mech Med Biol* 13 (2013): 1350067 [25 pages]. doi: 10.1142/S021951941350067X.

11. J. Buongiorno, Convective transport in nanofluids, *ASME J Heat Transfer* 128 (2006): 240–250.

12. D.A. Nield, A.V. Kuznetsov, The Cheng–Minkowycz problem for natural convection boundary-layer flow in a porous medium saturated by a nanofluid, *Int J Heat Mass Transfer* 52 (2009): 5792–5795.

13. M.M. Rashidi, O.A. Beg, N. Freidooni-Mehr, A. Hosseini, R.S.R. Gorla, Homotopy simulation of axisymmetric laminar mixed convection nanofluid boundary layer flow over a vertical cylinder. *Theor Appl Mech* 39 (2012): 365–390.

14. R.S.R. Gorla, A.J. Chamkha, A.M. Rashad, Mixed convective boundary layer flow over a vertical wedge embedded in a porous medium saturated with a nanofluid: Natural convection dominated regime, *Nanoscale Res Lett* 6 (207) (2011): 1–9.

15. P. Keblinski, S.R. Phillpot, S.U.S. Choi, J.A. Eastman, Mechanisms of heat flow in suspensions of nano-sized particles (nanofluid), *Int J Heat Mass Transfer* 45 (2002): 855–863.

16. S.K. Das, N. Putra, P.W. Thiesen, R. Roetzel, Temperature dependence of thermal conductivity enhancement for nanofluids, *J Heat Transfer* 125 (2003): 567–574.

17. M. Haghshenas Fard, M.R. Talaie, S. Nasr, Numerical and experimental investigation of heat transfer of ZnO/water nanofluid in the concentric tube and plate heat exchangers, *Therm Sci* 15 (2011): 183–194.

18. M. Mahmoodi, Mixed convection inside nanofluid filled rectangular enclosures with moving bottom wall, *Therm Sci* 15 (2011): 889–903.

19. S. Dinarvand, A. Abbassi, R. Hosseini, I. Pop, Homotopy analysis method for mixed convective boundary layer flow of a nanofluid over a vertical circular cylinder, *Therm Sci* doi: 10.2298/TSC1120225165D.

20. A.A. Abbasian Arani, M. Mahmoodi, S. Mazrouei Sebdani, M.A. Akbari, S. Nazari, Free convection of a nanofluid in a square cavity with a heat source on the bottom wall and partially cooled from sides, *Therm Sci* 18 (2014): 283–300.

21. M. Kalteh, A. Abbassi, M. Saffar-Avval, J. Harting, Eulerian–Eulerian two-phase numerical simulation of nanofluid laminar forced convection in a microchannel, *Int J Heat Fluid Flow* 32 (2011): 107–116.

22. S.E. Maige, C.T. Nguyen, N. Galanis, G. Roy, Heat transfer behaviors of nanofluids in a uniformly heated tube, *Super lattices Microstruct* 35 (3–6) (2004): 543–557.

23. G. Roy, C.T. Nguyen, P.-R. Lajoie, Numerical investigation of laminar flow and heat transfer in a radial flow cooling system with the use of nanofluids, *Super lattices Microstruct* 35 (3–6) (2004): 497–511.

24. K. Khanafer, K. Vafai, M. Lightstone, Buoyancy driven heat transfer enhancement in a two dimensional enclosure utilizing nanofluids, *Int J Heat Mass Transfer* 46 (2003): 3639–3653.

25. A. Akbarinia, A. Behzadmehr, Numerical study of laminar mixed convection of a nanofluid in horizontal curved tubes, *J Appl Thermal Eng* 27 (2007): 1327–1337.

26. A. Akbarinia, Impacts of nanofluid flow on skin friction factor and Nusselt number in curved tubes with constant mass flow, *Int J Heat Fluid Flow* 29(1) (2008): 229–241.

27. F. Talebi, A.H. Mahmoudi, M. Shahi, Numerical study of mixed convection flows in a square lid-driven cavity utilizing nanofluid, *Int Commun Heat Mass Transfer* 37 (2010): 79–90.

28. M. Shahi, A.H. Mahmoudi, F. Talebi, Numerical study of mixed convective cooling in a square cavity ventilated and partially heated from the below utilizing nanofluid, *Int Commun Heat Mass Transfer* 37 (2010): 201–213.

29. L.S. Sundar, K.V. Sharma, S. Parveen, Heat transfer and friction factor analysis in a circular tube with Al₂O₃ nanofluid by using computational fluid dynamics, *Int J Nanoparticles* 2(1–6) (2009): 191–199.

30. R. Lotfi, Y. Saboohi, A.M. Rashidi, Numerical study of forced convective heat transfer of nanofluids: Comparison of different approaches, *Int Commun Heat Mass Transfer* 37 (2010): 74–78.

31. M. Ishii, *Thermo-Fluid Dynamic Theory of Two-Phase Flow*, Eyrolles, Paris, France, 1975.

32. M. Manninen, V. Taivassalo, S. Kallio, *On the Mixture Model for Multiphase Flow*, VTT Publications 288, Technical Research Center of Finland, Espoo, Finland, 1996.

33. A. Akbarinia, R. Laur, Investigation the diameter of solid particles affects on a laminar nanofluid flow in a curved tube using a two-phase approach, *Int J Heat Fluid Flow* 30(4) (2009): 706–714.

34. S. Mirmasoumi, A. Behzadmehr, Effect of nanoparticles mean diameter on mixed convection heat transfer of a nanofluid in a horizontal tube, *Int J Heat Fluid Flow* 29 (2008): 557–566.

35. R. Mokhtari Moghari, A. Akbarinia, M. Shariat, F. Talebi, R. Laur, Two phase mixed convection Al₂O₃–water nanofluid flow in an annulus, *Int J Multiphase Flow* 37 (2011): 585–595.

36. V.D. Sakalis, P.M. Hatzikonstaninoi, N. Kafousias, Thermally developing flow in elliptic ducts with axially variable wall temperature distribution, *Int J Heat Mass Transfer* 45 (2002): 25–35.

37. K. Velusamy, V.K. Garg, G. Vailyanathan, Fully developed flow and heat transfer in semi-elliptical ducts, *Int J Heat Fluid Flow* 16 (1995): 145–152.

38. K. Velusamy, V.K. Garg, Laminar mixed convection in vertical elliptic ducts, *Int J Heat Mass Transfer* 39(4) (1996): 745–752.

39. L. Schiller, A. Naumann, A drag coefficient correlation, *Z Ver Deutsch Ing* 77 (1935): 318.

40. C.W. Ossen, Uber den goltigkeitsbereich der stokesschen widerstansformel, *Ark Mat Astron Fysik* 9 (1913): 33–43.

41. I. Proudman, J.R.A. Pearson, Expansions at small Reynolds number for the flow past a sphere and a cylinder, *J Fluid Mech* 2 (1957): 237–262.

42. K.A. Clift, W.H. Gauvin, The motion of particles in turbulent gas stream, *Proc Chem ECA* 1 (1970): 14–24.

43. N. Masoumi, N. Sohrabi, A. Behzadmehr, A new model for calculating the effective viscosity of nanofluids, *J Phys D Appl Phys* 42 (2009): 055501 (p. 6).

44. C.H. Chon, K.D. Kihm, S.P. Lee, S.U.S. Choi, Empirical correlation finding the role of temperature and particle size for nanofluid (Al_2O_3) thermal conductivity enhancement, *J Appl Phys* 87(5) (2005): 153107 (p. 3).

45. M. Shariat, R. Mokhtari Moghari, S.M. Sajjadi, M. Khojamli, Numerical investigation of Al_2O_3/water nanofluid in horizontal elliptic ducts using two-phase mixture model, *J Comput Theor Nanosci* 10 (2012): 1–9.

46. S.V. Patankar, *Numerical Heat Transfer and Fluid Flow*, Hemisphere, Washington, DC, 1980.

47. J. Schenk, B.S. Han, Heat transfer from laminar flow in ducts with elliptic cross section, *Appl Sci Res* 17 (1966): 96–114.

5

Nanooncology: Molecular Imaging, Omics, and Nanoscale Flow-Mediated Medicine Tumors Strategies

Tannaz Farrahi, Tri Quang, Keerthi Srivastav Valluru,
Suman Shrestha, George Livanos, Yinan Li, Aditi Deshpande,
Michalis Zervakis, and George C. Giakos

CONTENTS

5.1 Introduction 122
5.2 Fluid Dynamics in Tumors and Drug Transport 127
    5.2.1 Tumor Blood Flow 131
    5.2.2 Drug Transport 132
        5.2.2.1 Solute Dynamics in the Blood Vessel (Ω_v) 132
        5.2.2.2 Extracellular Drug Concentration 133
        5.2.2.3 Intracellular Drug Concentration 133
    5.2.3 Intracellular Apoptosis Signaling 133
        5.2.3.1 Bistable Switch 133
        5.2.3.2 Irreversible Monostable Switch 134
    5.2.4 Tumor Cell Density Dynamics 134
5.3 Nanotechnology-Based Drug Delivery in Clinical
    Pharmacokinetics 134
    5.3.1 Nanoparticles for Drug Delivery 136
    5.3.2 Lipid Nanoparticles 137
    5.3.3 Polymeric Nanoparticles 138
    5.3.4 Other Nanoparticles 139
5.4 Molecular and Omics Imaging and Spectroscopy for Cancer
    Detection 139
    5.4.1 QD-Based Detection of Primary Tumor 147
5.5 Paradigm: Nanoscale Flow Imaging 149
    5.5.1 Physical Principles 149
    5.5.2 Polarimetric Fusion with Statistical Models 150
        5.5.2.1 Models and Applications 150
        5.5.2.2 Computational Algorithm for Single- and
                 Multimodal Distributions 152

5.6   Contrast Discrimination Measures 154
    5.6.1   First Set of Experiments (L-Phenylalanine in
           Aqueous Solution at 830 nm) 155
    5.6.2   Second Set of Experiments (Glucose in Aqueous
           Solution at 655 nm) 157
    5.6.3   Therapeutics 161
5.7   Conclusion 162
References 162

5.1 Introduction

Cancer is a major threat for public health and a significant cause of mortality. In the battle against cancer, not only treatment but also early detection has paramount significance [1–27,31–34].

Nanooncology is the application of nanotechnology aimed at enhancing both the diagnostic and therapeutic content in the war against cancer. Combining nanotechnology and nanoscience principles together with molecular imaging, and *omics* disciplines, such as genomics, transcriptomics, proteomics, and metabolomics, nanooncology is poised to revolutionize the clinical arena; by introducing efficient and reliable methodologies for the diagnosis, assessment, monitoring, treatment, and management of cancer [6,10–12].

Fluid mechanics is a major contributing factor to the growth, progression, metastasis, and treatment of cancer. Oxygen and nutrients transport to cancerous tissues is mediated by the blood vessels that provide a route for metastasizing cancer cells to distant organs and deliver drugs to tumors. Cancer is the major threat of public health, and recent researches vastly improved the understanding of cancer and treatment. Early detection of cancer, besides prevention, is the most important factor in fight against cancer [1,6,10–12]. In the field of nanomedicine, nanooncology refers to the detection of biomolecules at the nanoscale level, such as metabolites, proteins, genes, and enzymes, as well as applications of nanotechnology in cancer detection, diagnosis, management, and therapy. Indeed, nanooncology has the potential of developing nanodevices that could be introduced to the body as an agent to identify and locate cancer cells. Typically, these devices would consist of a nanosensor and a storage volume that contains the therapeutic medications [39–42]. These nanodevices are generally controlled by a computer program and monitor the in vivo activities externally. The advantage of these nanodevices is early cancer detection by circulation through the body [10]. Similarly, metabolomics biomarkers for cancer detection and/or assessment of treatment efficacy are explored at preclinical level by means of animal and human cell cultures, followed by validation in tumor tissue or biofluids. In fact, cancer biomarkers can contribute significantly to the diagnostics and therapeutics of cancer by detecting early enough the onset of the disease, as well as by assessing in real time the drug response. Therefore, the expedited development of biomarkers aimed at improved diagnosis and treatment of cancer is of paramount significance [5,13,14,87,100–107].

Molecular imaging experienced explosive growth over the last few decades with a phenomenal growth in the area of clinical oncology. As result, we are closer to the application of imaging in the clinical management of cancer patients. Molecularly targeted imaging agents are expected to broadly expand the capabilities of conventional anatomical imaging methods. Molecular imaging will allow clinicians to provide both localization and visualization of the expression and activity of specific molecules, such as proteases and protein kinases, as well as biological processes, such as apoptosis, angiogenesis, and metastasis, all mechanisms that impact directly the tumor behavior and/or response to therapy. This information is expected to have a major impact on cancer detection, patient-dedicated treatment, and drug development, as well as improve our understanding of the cancer generation mechanisms and evolutionary processes [1,2,10,12,13,26,47,48,55, 60,61,87,104].

Metabolomics is a clinical science field aimed at the global quantitative assessment of endogenous metabolites within a biological system. Metabolomics provides an attractive platform to assess hundreds of metabolites simultaneously and in an unbiased manner. Detection and identification of metabolites is carried out in cells, tissues, or biofluids using nuclear magnetic resonance (MR) and spectroscopy/liquid chromatography (MS-LC), which are the two major techniques used widely in metabolomics. Metabolomics, when used as a translational research tool, can provide a link between the laboratory and the clinic, particularly because metabolic and molecular imaging technologies, such as positron emission tomography and MR spectroscopic imaging, enable the discrimination of metabolic markers noninvasively in vivo [1,5,7,12,40,46,50,54–61,112–116,118,119].

Nanotechnology, the engineering of functional systems at the molecular scale, is a powerful and potential tool with vast applications in medicine, electronics, biomaterials, and energy production. In recent years, several biomedical researches based on nanotechnology have been performed, resulting in significantly enhanced efficacy in both diagnostics and therapeutics [2,3,8–11,16].

Nanotechnology offers a new set of tools for cancer detection and diagnosis. One example is the cancer biomarker. According to the U.S. National Cancer Institute, biomarker is defined as "a biological molecule found in blood, other body fluids, or tissues that is a sign of a normal or abnormal process, or of a condition or disease" [2,3,5,8–28,31,111].

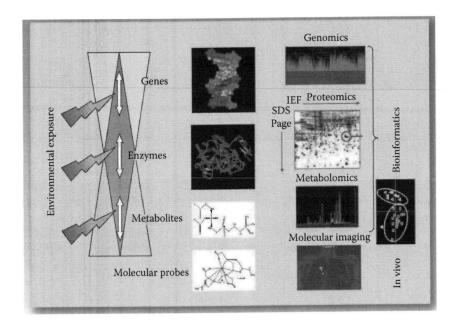

FIGURE 5.1
(See color insert.) Molecular imaging and omics technologies. (From Spratlin, J.L. et al., *Clin. Cancer Res.*, 15(2), 431, 2009.)

Molecular imaging and *omics* technologies are depicted in Figure 5.1 [60]. Cancer biomarker is the molecular signature that indicates the physiological changes of tissue or cell during cancer progress and can be used to see how body responds to therapeutic interventions [40,112–116].

These biomarkers are so beneficial for cancer diagnosis and continuous cancer monitoring. Nanotechnology goes beyond the molecular diagnosis of cancer by using nanoparticles such as gold nanoparticles and quantum dots (QDs). Application of QDs in cancer diagnosis has a wide interest compared with previous fluorescent markers. QDs are considered as inorganic fluorophores, which are tunable based on their size, thereby having the ability to have a wide range of optical properties; for example, they provide broader excitation spectra than organic fluorophores, and their excitation spectra contain ultraviolet (UV) to red. Gold nanoparticles have few advantages over QDs, which use semiconductor materials. Gold nanoparticles are not toxic and can be used inside the human body [1]. The scattering of gold nanoparticles is so strong and can be observed using white light and a simple, inexpensive microscope. Nanoparticles can serve as nanoagents in imaging of tumors and improve the specificity and sensitivity of MR imaging [2]. On the other hand, nanoparticles can be used as imaging contrast agents.

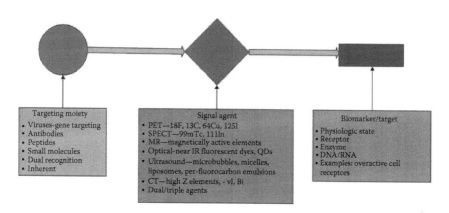

FIGURE 5.2
(See color insert.) The impact of imaging in the evolutionary process of cancer.

Conventional imaging like magnetic resonance imaging (MRI), computer tomography (CT), and ultrasound used for cancer detection usually has more efficacies when cancer becomes a mass. Utilizing nanoparticles as imaging agents provides molecular imaging, which has the potential to detect cancer at early stages and provide in vivo characterization information for predicting the most beneficial therapy for the patient. In optical imaging, nanoparticles produce high-intensity signals and image fewer cells at specific depth and produce more stable signals over a longer period of time [3]. The impact of imaging in the evolutionary process of cancer [1] is shown in Figure 5.2.

On the other hand, the evolutionary process of cancer is shown in Figure 5.3. Early diagnosis has been the major factor in the reduction of mortality and cancer management costs.

The term *theranostics* was coined to describe the ability of a single nanoagent to provide both therapeutic and diagnostic functions. In other words, its aim is to have an agent, for example, a nanoparticle, with the ability of simultaneous diagnosis and treatment. The idea is to develop a smart nanoparticle to detect, deliver the cancer medication to the cancerous cells, and monitor the response of the cells to the therapy. Nanoparticle-based imaging and therapy has been studied separately, but the idea of nanoparticle-based theranostics is to focus more on codelivery in addition to previous therapeutic function, which means delivering imaging and therapeutic functions simultaneously during the treatment [12,17–24,29–30,110]. We will discuss more about the function of nanoparticles in early detection of diseases and as therapeutics agents, carriers of chemotherapeutical drugs to assist the accurate delivery of drugs, targeting tumors and photothermal therapy of tumors. Interestingly enough, over the last decades, there has been a wide research and development on nanotechnology platforms to diagnose and treat the cancer [17]. Some of the nanoplatforms are liposomes, polymer–drug conjugates, and dendrimers.

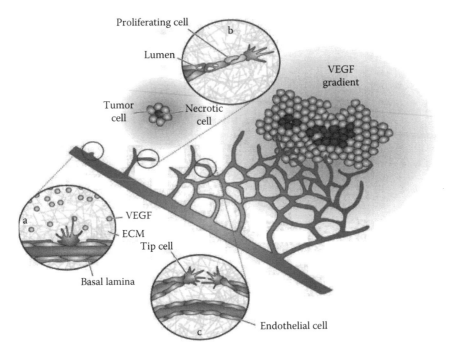

FIGURE 5.3
(See color insert.) Steps involved in promoting angiogenesis in tumors. (From Koumoutsakos, P. et al., *Annu. Rev. Fluid Mech.*, 45, 325, 2013.)

Compared with conventional drug delivery systems, nanomedicine plat-forms provide multiple functioning and tend to become a new paradigm for the detection, treatment, and monitoring of cancer [17]. Nanoscale materi-als have unique chemical and physical properties such as small size, in the range of 1–1000 nm [9], large surface area to mass ratio, and high reactivity and can interact with cells at the molecular level. In terms of therapeutics, there are several potential approaches for cancer therapy such as photo-dynamic therapy (PDT), photothermal therapy, nanobody-based cancer therapy, nanobomb, and drug delivery. One of the major applications of nanotechnology in cancer therapeutics is drug delivery. The effectiveness of an anticancer drug is not only determined by the drug itself but also the ability of drug delivery to cancer tissue. The significances of effective drug delivery in cancer take into account the optimal therapeutic effect of drugs and considerable cutback of toxic side effects. Typically, the full therapeutic effect of drugs in cancer is limited by drug resistance due to noncellular mechanisms impeding drug diffusion across the cell membrane and cellular mechanisms referring to biochemical and metabolic changes inside cancer cells, and intracellular and extracellular transport mechanisms. Moreover, lack of specificity of chemotherapeutic drugs results in systemic toxicity, causing significant damage to normal tissues and lower concentrations of drug delivered to the tumor [12]. Currently, novel pharmaceutical formulations have been developed and utilized to design efficient drug delivery systems in which the drug is distributed to the tumor at higher concentrations with its minimal exposure to noncancerous tissues. Approximately 150 drugs in the development of cancer are based on nanotechnology [1]. Fundamentally, the surface area to volume ratio relevant to functional surface area is an important factor for the reactivity of a material. Nanoparticles have a large functional surface area due to their nanoscale size that allows them to interact with biological systems at the cellular level. Therefore, anticancer drug particles can be incorporated, entrapped, or encapsulated into nanoparticles, and then transported to the cancer cells. Upon the tunability in structure and properties, nanoparticles, moreover, have remarkable advantages such as high stability, biocompatibility, and effective transportability during the drug delivery process. All unique biological, chemical, and physical properties are employed to functionalize nanoparticles as nanoscale carriers or nanoparticle drug complexes for nanotechnology-based drug delivery in clinical pharmacokinetics. Incorporation of chemotherapeutic drugs into nanoparticles also creates pharmaceutical formulations at the nanoscale level.

5.2 Fluid Dynamics in Tumors and Drug Transport

A physical algorithm of the etiology and evolutionary progress of cancer, monitoring, and treatment is missing, although tremendous efforts are made in integrating physics, optics, and mechanics with genomic research of cancer. Fluid mechanics is a major contributing factor in the growth, progression, metastasis, and treatment of cancer. Blood flow has a dual role in many forms of carcinogenesis: (a) it provides oxygen and nutrients to tumors and (b) determines the routes of metastasizing cancer cells through flow patterns in blood and lymphatic vessels. With respect to point (a), metabolic activity inside the tumor requires sustained amounts of oxygen and nutrients, which may be provided by diffusion through the surrounding perfused tissue [6,14,15,45].

In this so-called avascular stage, tumor volumes usually do not exceed 1 mm^3. Tumor angiogenesis reflects several phenotypic variations. Low- or high-density angiogenesis is present depending upon the type and location of the tumors. The steps involved in promoting angiogenesis in tumors, which is characterized by low shear stress conditions inside the vessel and a positive VEGF gradient, are shown in Figure 5.3.

Indeed, fluid mechanics processes are active participants in lymphatic metastasis [6,44].

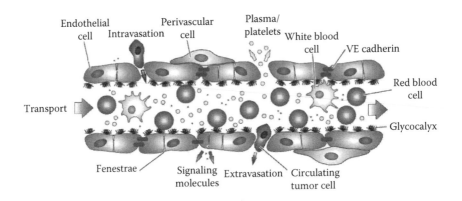

FIGURE 5.4
(See color insert.) Tumor blood vessel structure, intravasation, transport, and extravasation of the blood constituents. (From Koumoutsakos, P. et al., *Annu. Rev. Fluid Mech.*, 45, 325, 2013.)

The morphology and structure of the vasculature inside tumors is different from the vasculature observed in normal tissue. Specifically, tumor vasculature exhibits pronounced density and ramification patterns, enlarged distorted vessels, and highly tortuous, abnormal endothelial cell (EC) stratification, often dead-end segments, altered basement membranes, large gaps between ECs of the vasculature, and blood flow irregularities. The internal tumor vessel structure, intravasation, transport, and extravasation of the blood constituents are shown in Figure 5.4 [6].

Typically, cell transits from its physiological state to a cancerous state through the accumulation of a set of mutations in its genome.

The tumor microenvironment consists of the surrounding tissue, extracellular matrix (ECM), with abundant soluble growth factors. Fibroblast cells, key constituents of the tumor microenvironment, release the ECM molecules that compose the tumor stroma. Typically, these structures are altered by cancer cells and their interaction with the microenvironment, leading to an increase in the observed stress, such as solid radial stresses in the tissue surrounding the tumor and increased stress inside the tumor, a stiffened ECM, due to elevated collagen, elevated fluid pressure, contraction during the rearrangement of the ECM fibers, and interstitial flow. These mechanisms facilitate the metastasis of cancer cells.

The landscape of the tumor microenvironment is shown in Figure 5.5 [6].

The transport of anticancer drugs to tumor cells and their responses to the administrated drug [1] have a major effect on the efficiency of chemotherapy. Several mathematical modeling frameworks have been developed on tumor drug transport. In those studies, tumor vasculature is treated as a distributed source [99–103]. Considering the irregular and heterogeneous nature of the vasculature geometry, realistic assumptions are needed. In another study, the effects of glycolysis on tumors (brain gliomas) at both the genomic and metabolic level were considered [104]. An integrated systems-based modeling

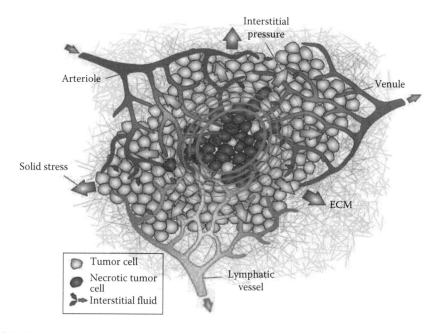

FIGURE 5.5
(See color insert.) Landscape of the tumor microenvironment. (From Koumoutsakos, P. et al., *Annu. Rev. Fluid Mech.*, 45, 325, 2013.)

framework for drug transport and its effect on tumor cells has been recently reported [105].

In this model, basic descriptions of blood flow, drug transport, intracellular apoptosis signaling, and tumor cell density dynamics are included. It relies on the four following assumptions [105]:

1. Blood is an incompressible Newtonian fluid, while the blood vessel is straight and rigid, ignoring tortuous shape.
2. The tumor interstitium is homogeneous, ignoring heterogeneity, with a uniform distribution of nutrients and pH.
3. Tumor cells are stationary, leading to the assumption that the tumor interstitium has a fixed outer boundary.
4. All tumor cells are distributed uniformly, identical, that is, ignoring cellular variability, stochasticity, and the effects of cell-to-cell interactions and cell-cycle, and alive initially.

The information flow included in the modeling framework is shown in Figure 5.6.

The equations governing the modeling framework of [6] are summarized in Figure 5.7, where the definitions of the symbols in the modeling framework are offered in Table 5.1.

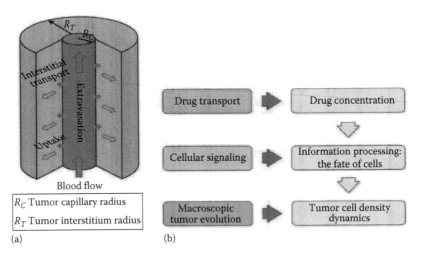

(a)   (b)

FIGURE 5.6
(See color insert.) Information flow of the modeling framework. (a) Simplified diagram of the tumor vascular geometry and (b) flow chart of the algorithm included in the model. (From Koumoutsakos, P. et al., *Annu. Rev. Fluid Mech.*, 45, 325, 2013.)

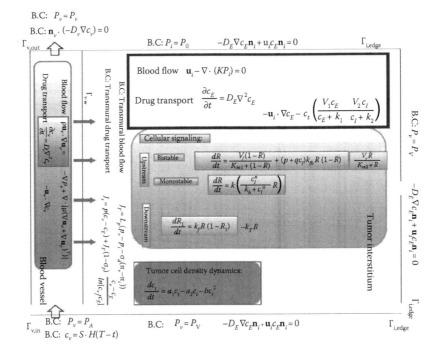

FIGURE 5.7
Equations governing the modeling framework. (From Koumoutsakos, P. et al., *Annu. Rev. Fluid Mech.*, 45, 325, 2013.)

TABLE 5.1

Definitions of the Symbols Used in the Modeling Framework

Symbol	Definition
u_v	Blood velocity vector in vascular domain
u_i	Blood velocity vector in interstitial domain
P_v	Vascular blood pressure
P_i	Interstitial fluid pressure
J_F	Transmural fluid velocity, determined by Starling's law
J_s	Transmural drug flux, determined by Kedem–Katchalsky
c_v	DOX concentration in vascular domain
c_E	Extracellular DOX concentration
c_I	Intracellular DOX concentration
R	Hypothetical downstream intermediate protein in apoptosis signaling cascade
R_1	Hypothetical protein responsible for triggering apoptosis
c_t	Tumor cell density

The mathematical framework addresses the following areas:

1. Tumor blood flow
2. Drug transport
   a. Drug concentration in the vascular space
   b. Intracellular drug concentration
   c. Intracellular apoptosis signaling
3. Intracellular apoptosis signaling
4. Tumor cell density dynamics

5.2.1 Tumor Blood Flow

By means of the Navier–Stokes equations:

$$\rho u_v \cdot \nabla u_v = -\nabla P_v + \nabla \cdot \left[\mu \left(\nabla u_v + \left(\nabla u_v \right)^T \right) \right] \qquad (5.1a)$$

$$\nabla \mathbf{u}_v = 0 \qquad (5.1b)$$

where
ρ is the blood density
μ is the blood viscosity
P_v is the vascular blood pressure

Solving Equations 5.1a and 5.1b to the boundary conditions (5.2a) through (5.2c):

$$P_v = P_A \quad \text{on } \Gamma_{v,in} \tag{5.2a}$$

$$P_v = P_V \quad \text{on } \Gamma_{v,out} \tag{5.2b}$$

$$\mathbf{n}_v.\mathbf{u}_v = J_F \quad \text{on } \Gamma_{v-i,w} \tag{5.2c}$$

The transmural velocity, J_F, can be calculated using Starling's law:

$$J_F = L_P \left(P_v - P_i - \sigma_d \left(\pi_v - \pi_i \right) \right) \quad \text{on } \Gamma_{v-i,w} \tag{5.3}$$

where
L_P is the vascular hydraulic conductivity
σ_d is the osmotic reflection coefficient
π_v and π_i are osmotic pressure in the vascular and interstitial space, respectively

Since $(\pi_v - \pi_i) \cong 0$ in solid tumors

$$J_F = L_P \left(P_v - P_i \right) \quad \text{on } \Gamma_{v-i,w} \tag{5.4}$$

5.2.2 Drug Transport

The intravascular drug concentration (c_v), the interstitial extracellular free drug concentration (c_E), and intracellular drug concentration (c_I) had been taken into consideration in order to describe the drug transport.

5.2.2.1 Solute Dynamics in the Blood Vessel (Ω_v)

It is expressed by means of Equation 5.5 as

$$\frac{\partial c_v}{\partial t} = D_v \nabla^2 c_v - \mathbf{u}_v \cdot \nabla c_v \tag{5.5}$$

where
c_v refers to the drug concentration in the vascular space
D_v is the drug diffusivity in the vascular space

The solute dynamics in the interstitium (Ω_i) is expressed in terms of the extracellular and intracellular drug transport.

5.2.2.2 Extracellular Drug Concentration

$$\frac{\partial c_E}{\partial t} = D_E \nabla^2 c_E - \mathbf{u}_i \cdot \nabla c_E - c_t \left(\frac{V_1 c_E}{c_E + k_1} - \frac{V_2 c_I}{c_I + k_2} \right) \tag{5.6}$$

where

c_E and c_I refer to the extracellular and intracellular drug concentration, respectively

D_E is the diffusion coefficient of drug in the interstitium

c_t is the tumor cell density

V_1, V_2, k_1, and k_2 are constants that describe transport across the cell membrane, in which V_1 and V_2 are the maximum concentration rates of transmembrane transport and k_1 and k_2 are the Michaelis–Menten constants for transmembrane transport.

5.2.2.3 Intracellular Drug Concentration

There is an interrelationship between the intracellular drug concentration and the extracellular drug concentration according to transmembrane transport, offered as

$$\frac{\partial c_I}{\partial t} = \frac{V_1 c_E}{c_E + k_1} - \frac{V_2 c_I}{c_I + k_2} \tag{5.7}$$

5.2.3 Intracellular Apoptosis Signaling

The apoptosis models are based on the two types of switches, commonly observed in cellular signaling: bistable and monostable apoptosis switches [6].

5.2.3.1 Bistable Switch

$$\frac{dR}{dt} = \frac{V_f(1-R)}{K_{m1}+(1-R)} + (p+qc_I)k_{fb}R(1-R) - \frac{V_r R}{K_{m2}+R} \tag{5.8}$$

where

K_{m1} and K_{m2} are the Michaelis–Menten parameters

k_{fb} is a kinetic parameter that parameterizes the feedback strength

The constants p and q serve to set the basal level and dynamic range of the module

5.2.3.2 *Irreversible Monostable Switch*

$$\frac{dR}{dt} = k\left(\frac{c_l^n}{k_h + c_l^n} - R\right) \tag{5.9}$$

where
  n denotes the Hill coefficient
  k_h is an associated constant in the Hill term
  k is a parameter representing the time scale of the response

5.2.4 Tumor Cell Density Dynamics

$$\frac{dc_t}{dt} = a_1 c_t - a_2 c_t - bc_t^2 \tag{5.10}$$

where
  a_1 is the tumor growth rate
  a_2 is the tumor natural decay rate
  b is the saturation constant in the tumor growth equation

5.3 Nanotechnology-Based Drug Delivery in Clinical Pharmacokinetics

Drug resistance caused by noncellular and cellular mechanisms significantly restricts drug delivery to cancer cells, thereby reducing the therapeutic effect. Besides, lower concentration of cytotoxin drugs directly delivered to cancerous cells increases systemic toxicity, which causes severe side effects with the treatment. Nanooncology is an approach that revolutionizes the development of novel pharmaceutical formulations and advanced technologies for efficient drug delivery systems, thus yielding the optimal therapeutic effect in cancer. Nanotechnology-based drug delivery systems comprise two main components, namely, anticancer drugs and nanoparticles. Drugs can be divided into two types—hydrophobic and hydrophilic drugs—based on their surface characteristics. In addition to the advantages of large functional surface areas, tunability in structure and properties, high stability, biocompatibility, and effective transportability, nanoparticles function as nanoscale carriers or nanoparticle drug complexes in order to deliver and target both hydrophobic and hydrophilic drugs to tumors.

A general process of the nanotechnology-based drug delivery in nanooncology is shown in Figure 5.8. In order to perform the drug transportation function, the nanoscale carriers first bind drug molecules on their surfaces (e.g., metal nanoparticles), encapsulate (e.g., liposomes), or entrap

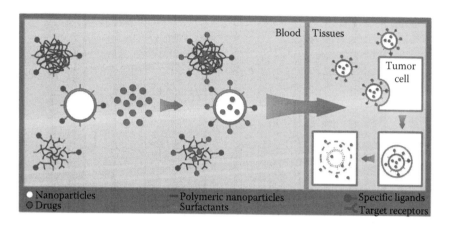

FIGURE 5.8
(See color insert.) A schematic illustration of nanotechnology-based drug delivery systems. (From Thakor, A.S. and Gambhir, S.S., *CA Cancer J. Clin.*, 63(6), 395, 2013.)

(e.g., polymeric nanoparticles) drugs based on their specific structures. The combination between anticancer drugs and nanoscale carriers forms a nanoparticle drug complex in which anticancer drugs are confined and protected from quick degradation, thereby retaining their concentration and pharmaceutical properties in the blood circulation as well as preventing damages from systemic toxicity. Commonly, foreign substances such as injected nanoparticle drug complexes tend to be removed from the body quickly through renal excretion, or cleared by the reticuloendothelial system (RES) because of the body's defense mechanisms. Therefore, surfactants such as polyethylene glycol (PEG) and other hydrophilic conjugates are exploited to coat the nanoparticles so as to improve their stability and biocompatibility and reduce the clearance time, allowing drugs to stay longer in the blood circulation. Consequently, the prolonged acting effect of drugs allows longer dosing intervals and decreased cytotoxicity concurrently [12].

Nanoparticle drug complexes are then injected into the systemic circulation and flow within the bloodstream to be delivered into cells. Nanoparticles have the capability to target specific tumor cells based on the enhanced permeability and retention (EPR) effect of tumors and nanoparticle surface modifications. The EPR effect refers to the passive drug delivery in which more blood supply to tumor tissues upon their abnormal architecture in the process of angiogenesis and lack of lymphatic drainage leads to an increased drug delivery and accumulation in the tumor mass in comparison with normal tissues. Detailed explanations of EPR effect can be found in [12]. However, the EPR effect is limited by the tumor blood flow heterogeneity and the impediment of noncellular mechanisms on drug diffusion across the cell membrane [70]. Alternatively, the active drug delivery referring to nanoparticle surface modifications has been developed in recent researches, providing an enhanced targeted drug delivery to tumors. For actively targeting tumor cells, nanoparticles are modified by conjugating ligands to the surfactants on their surface. The ligands are able to recognize and bind to one or multiple specific receptors that are present at the surface of tumor cells.

After that nanoparticle drug complexes delivered to the tumor site through the bloodstream, they penetrate into tumor cells via receptor-mediated endocytosis and then disincorporate to release the drugs. Many recent studies have presented novel approaches that allow drug release to be triggered as nanoparticles interact with the UV light [71], the tumor microenvironment [72–74], or hyperthermia [75]. The reduced size of nanoparticles upon illumination of UV light or proteases in tumor microenvironment and changes in the configuration of hybrid nanoparticles developed with many layers that differently respond to changes in tumor microenvironment conditions such as oxidative stress, temperature, or pH allow the nanoparticles penetrate deeper and release drugs into tumor tissues. A novel paradigm exploits mild hyperthermia to act as a tumor-localized release trigger for the low-temperature-sensitive liposome (LTSL). This temperature trigger allows anticancer drug to release rapidly by using microwaves, radiofrequency (RF), or ultrasound to heat tumors to the transition temperature of LTSL, optimally at 41.3°C [75]. Drugs released outside can enter tumor cells either through active transport mechanisms or via receptor-mediated endocytosis. However, all of the nanoparticle drug complexes are supposed to release drugs after they enter tumor cells for the optimal drug concentration and pharmaceutical properties and minimum side effects from systemic toxicity. Thus, the spatiotemporal control of drug release is very essential for an efficient and safe drug delivery system. The diffusion of drugs out of their matrix can be controlled and sustained as confined and protected by nanoparticles. Some nanoparticles made of biodegradable materials allow drugs to release after their degradation. As soon as they enter into the tumor cells, the release process of drugs can be hastened by taking advantages of biochemical reactions between nanoparticles and available agents in the cytoplasm.

The removal of residual nanoparticles after drug release is one of the leading concerns in the development and fabrication of nanocarriers. Nanoparticles should have degradability within a sufficient lifetime that allows the nanocarriers to be stable until drugs are delivered into tumor cell. Alternatively, materials lacking biodegradation such as inorganic nanoparticles that show an outstanding stability during drug delivery process must have the capability to dissolve into nonharmful components and then be safely metabolized or removed out of the body.

5.3.1 Nanoparticles for Drug Delivery

Two basic methods for the synthesis of nanostructure presented in [76] are *bottomup* and *topdown* approaches. In the *bottomup* approach, atoms or molecules are assembled by utilizing controlled chemical reactions, minute probes, and templating and nontemplating techniques to form nanoparticles. On the other hand, the *topdown* approach exploits bulk or film, surface, and mold machining based on lithography to break, cut, or etch bulk materials for the achievement of nanostructures. Developments of nanoparticle drug complexes in which one or more anticancer drugs are incorporated into nanoparticles as a nanoscale drug carrier come up with novel drug formulations. In general, the purpose of this combination is to employ the advantages of nanoparticles to maintain drug concentration and pharmaceutical properties in blood circulation, prevent damages from systemic toxicity, and thus enhance the effect of drug delivery. In nanooncology, several formulations have been developed and widely applied in varieties of cancers [110].

5.3.2 Lipid Nanoparticles

Liposome as an organic particle has a self-assembling structure composed of an aqueous core surrounded by a concentric lipid bilayer. Possessing both hydrophilic and lipophilic properties enables liposomes to encapsulate hydrophilic drugs in the interior aqueous space and entrap lipid-soluble drugs between the hydrophobic tails of the phospholipids. The advantages of liposomes are the ability to enhance the solubility of drugs such as paclitaxel and camptothecin [70] and biodegradability to be safely removed after drug release. Typically, doxorubicin, approved by the U.S. Food and Drug Administration (FDA) as a common anticancer drug, is formulated to liposomal formulations such as Doxil and Myocet in use for the treatment of metastatic breast, metastatic ovarian, bladder, lung cancer, and AIDS-related Kaposi sarcoma [11]. A novel method has recently highlighted the potential of thermally sensitive liposomes (Dox-TSLs) in fast release of doxorubicin inside the tumor vasculature when Dox-TSLs are heated to their transition temperature in the 40°C–42°C range [77]. One drawback of liposomes is their rapid clearance from the blood circulation by phagocytic cells of the RES. Surface modifications by binding surfactants such as glycolipids, PEG, or other hydrophilic conjugates are utilized to significantly improve biocompatibility, stability, and blood circulation half-life [12]. For example, the half-life of Myocet is prolonged to 2.5 h, and Doxil is 55 h while free doxorubicin is limited to 0.2 h [78]. Alternatively, solid lipid nanoparticles (SLNs) formulated with a rigid core of hydrophobic lipids surrounded by a monolayer of phospholipids show better stability compared with liposomes. Nevertheless, limitations in the drug pay loading capability and drug expulsion are still challenging in the design of SLNs. Lipid-based nanoparticles are modified with appropriate conjugating ligands to improve their selectivity and specificity for an enhanced targeted drug delivery. A schematic illustration of a liposomal formulation is shown in Figure 5.9.

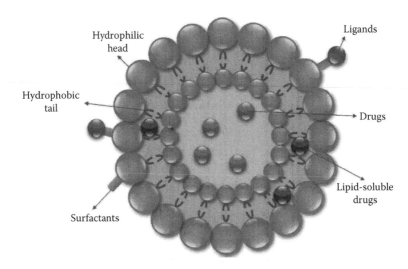

FIGURE 5.9
(See color insert.) A schematic illustration of a liposomal formulation.

5.3.3 Polymeric Nanoparticles

Over the last decades, polymeric nanoparticles have attracted the tremendous interest of biomaterial researches in nanooncology. Basically, polymeric nanoparticles include nanospheres and nanocapsules with a size range of 10–1000 nm [11]. Nanospheres with a spherical and solid structure can absorb drugs to their surface while nanocapsules have the ability to encapsulate drugs in their water or oil core surrounded by a solid shell. Dispersion of preformed polymers and polymerization of monomers are the main methods used for polymeric nanoparticle preparation [79]. Polymeric nanoparticles consist of synthetic polymers such as poly(ε-caprolactone), polylactic acid (PLA), poly(lactide-coglycolide) (PLGA), polyglycolic acid (PGA), PEG, and poly(alkylcyanoacrylate) and natural polymers such as gelatin, dextran ester, and chitosan and alginate. Generally, polymeric nanoparticles have higher stability, higher loading capacity for poorly water-soluble drugs, and greater potential in surface modifications. Typically, Abraxane formulated by paclitaxel bound to a natural polymeric nanoparticle made of albumin was approved by the FDA in 2005 for metastatic breast cancer cases. Abraxane takes advantage of albumin for an effective delivery of paclitaxel to tumors by both passive and active targeting manners. An example of efficient targeted codelivery of docetaxel and doxorubicin by aptamer nanoparticles with controlled release profiles based on differential drug release kinetics has also presented substantial potential for multifunctional polymeric nanoparticles [12,80].

Polymeric micelles have a core–shell structure with a size range of 10–200 nm composed of amphiphilic block copolymers that are self-assembled in an aqueous solution. The inner core formed by the hydrophobic polymers has a high loading capability of hydrophobic drugs. Consequently, polymeric micelles help prevent side effects from toxic adjuvants that are normally used to solubilize hydrophobic drugs. On the other hand, the outer shell formed by hydrophilic polymers exhibits an increase in stability and an active functional surface for further modifications. Substantially, polymeric micelles are degradable and capable of containing and codelivering many drugs at the same time. In addition, the dissociation and drug release of polymeric micelles can be regulated by change in pH, temperature, ultrasound, and lights [81]. Genexol-PM as novel Cremophor EL-free polymeric micelle formulation of paclitaxel (Taxol) designed with a hydrophobic inner core composed by poly(D,L-lactide) and a hydrophilic outer shell composed by PEG demonstrates significant antitumor efficacy with no additional toxicity for the treatment of metastatic breast, nonsmall-cell lung, and advanced pancreatic cancers compared with free paclitaxel [11].

5.3.4 Other Nanoparticles

Dendrimers have a spherical, highly branched, and synthetic structure that is composed of an initiator core, multiple layers of branched repeating units, and active end groups. The core of dendrimers is also called generation zero, and each layer is a generation. The size of dendrimers in the range of 1–15 nm and also their shape are adjustable. Besides these nanoparticles, other nanoparticles broadly used as nanoscale carriers in drug delivery comprise nanoshells, carbon oxide nanoparticles, glucose-based nanoparticles (dextran, chitosan, etc.), hydrogel nanoparticles, viral nanoparticles, polysaccharide-based nanoparticles, metallic nanoparticles, ceramic nanoparticles, and carbon nanotubes. Indeed, gold nanoparticle can be considered as a theranostic agent [1,9–12,15–17].

5.4 Molecular and Omics Imaging and Spectroscopy for Cancer Detection

The urgency for early diagnostics as well as effective treatment of complex diseases such as cancer, constantly increases the pressure on the development of efficient and reliable methods for targeted drug/gene delivery as well as molecular imaging techniques for diagnostics and theranostics. One of the most recent approaches covering both the drug delivery and the imaging aspects is benefitting from the unique properties of nanoscale materials. Nanoparticles, including fluorescent semiconductor nanocrystals (QDs) and magnetic nanoparticles, have proven their excellent properties for in vivo imaging techniques in a number of modalities such as MR, nuclear medicine, and fluorescence imaging. Specifically, nanoparticles have an advantage for molecular imaging in the sense that many functionalities can be added to the surface and interior of the particle. Nanoprobes for positron emission tomography (PET) of angiogenesis and cancer are one of the many examples.

Nanotechnology offers a new set of tools for molecular imaging of cancer. In the last few years, there has been a significant development of nanoparticle-based diagnostic and therapeutic agents for cancer. For diagnostic applications, nanoparticles provide the possibility of cancer detection at a molecular scale. They allow the detection of cancerous cells and biomarkers [1]. The impact of nanotechnology in the imaging process appears to be catalytic for the development of high-contrast, high-specificity molecular imaging techniques. Nanoparticles have several advantages over the traditional imaging agents such as higher signal intensity, more stable signals, enhanced specificity, improved aging characteristics (inorganic nanoparticles vs. proteins), and higher depth of penetration [12]. For instance, in an experiment, Rabin et al. injected polymer-coated Bi2S3 nanoparticle as an imaging agent into a mice. The results exhibited high x-ray absorption compared with iodine, longer time period of circulation in vivo (more than 2 h), and perfect stability in high concentrations and safety comparable to or better than iodinated molecules. Nanoparticles accumulated in lymph nodes that are metastatic show up as bright dots in CT images. Enhanced in vivo imaging of liver, vasculature, and lymph nodes of the mice is reported by Rabin et al. [1,47].

QDs are fluorescent nanocrystals that are made of semiconductor materials, and their physical dimensions are smaller than the exciton Bohr radius [18]. Because of their advantages over conventional fluorescent markers, there is an excessive interest in deploying QDs as inorganic fluorophores for molecular imaging applications. For instance, one of these advantages is their similarity with macromolecules dimensions. QDs have unique optical and electrical properties such as [9,10,18,20,21]

- Ultra small size, 2–10 nm in size
- Wavelength tunability based on their size and composition of the shell and core
- High sensitivity due to high quantum yield
- Photobleaching resistivity that is suitable for long-term imaging
- Large surface-to-volume ratio
- Narrow emission spectra and large absorption spectra that make them feasible to detect multiple signals with the same excitation, which is called multiplexing [23]

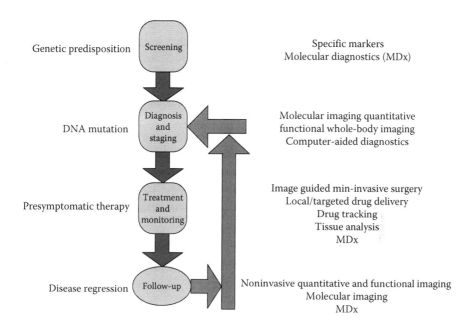

FIGURE 5.10
The impact of molecular imaging in the evolutionary process of cancer.

The impact of molecular imaging in the evolutionary process of cancer is shown in Figure 5.10.

Traditionally, imaging systems have provided clinicians and researchers noninvasive methods for the observation of internal bodily structures, determination of functional tissue characteristics, metabolic function, and identification of diseases and conditions. As technology advances, short data acquisition time, reduced cost, high spatial resolution, high contrast resolution, and high specificity images at a decreased radiation dose are realized, offering patients efficient diagnosis and decreased morbidity.

During the last few decades, new researches on imaging modalities changed the paradigm of imaging from anatomical imaging to molecular imaging and emerged the ability of detecting the cancer cells at the early stages. Molecular imaging is of paramount significance in predicting the best appropriate treatment for specific patient by allowing the in vivo characterization of genetic changes in oncogenesis.

For the sake of clarity, human anatomical and geometrical information can be displayed by digital radiography and CT imaging systems. Here, the transmission of ionizing radiation through the body and calculation of its subsequent attenuation by tissues and organs comprise the basis for which these systems operate. Both digital radiography and CT rely on an x-ray tube that generates a beam of ionizing radiation and slot-beam arrays or a flat panel of high-resolution, direct- or indirect-based ionization of photon electronic detectors, coupled to data acquisition electronics and signal conditioning circuitry. Typically, high atomic number, high density, high quantum efficiency amorphous silicon (a-silicon), ceramic scintillators, cadmium tungstate ($CdWO_4$), or cadmium zinc telluride (CZT) substrates are the detector materials of choice. However, these imaging systems do not offer physiological or metabolic signatures but rather structural and anatomical information.

On the other hand, PET, SPECT, ultrasound, and MRI demonstrate great clinical value for molecular and functional imaging over a wide range of diagnostic applications, although the true potential of each system within the clinical arena is still largely unexplored.

The roots of molecular imaging are with nuclear medicine as it relies on the emission of radiation by radiotracers from within the body. SPECT for molecular imaging requires the injection of gamma-ray-emitting isotopes, such as 99mTc, 111In, 123I, or 131I. Molecular imaging using PET can be achieved by labeling a natural biological molecule with a positron-emitting isotope, such as 14O, 13N, 11C, or 18F, providing accurate chemical classification. These labeled molecules are injected into the patient who is subsequently imaged. The sensitivity of PET, being 10^{-11}–10^{-12} mol/L, is an order higher than that of SPECT and is also independent of the depth of the probe in the tissue. MR imaging generates tomographic images by exploiting the nuclear MR properties of tissues, stimulated by applying fixed magnetic and varying RF fields. MR highlights anatomical structures, but functional MR is becoming more common. MR is becoming a very important noninvasive imaging modality since it provides unique contrast between soft tissues and very high spatial resolution. Yet the chief advantage of MR over other imaging modalities is the fact that it does not employ ionizing radiation. Coupled with contrast-enhancing, radiolabeled, and magnetically active imaging agents, these modalities provide clinicians with molecular information and insight into biological and pathological disease processes, resolving limitations introduced by traditional in vitro and histopathological methods. Ultrasound, MR, and CT imaging fundamentally rely on the ability to differentiate the target against the surrounding tissue and inherent background noise. As a result, they can produce signals with little sensitivity or specificity.

Each clinical modality highlights different physical parameters and functions of the human body and hosts different resolution and sensitivity characteristics. However, the appearance of multifusion-based modalities, for instance, PET-MRI or CT-SPECT-PET, looks extremely promising to the clinical community, leading to either further enhancement of the metabolic diagnostic content or simultaneous acquisition of anatomical, functional, and metabolic information, respectively. The contrast mechanisms offered by the different imaging modalities are given in Table 5.2.

Similarly, optical imaging offers the potential for noninvasive exploration of molecular targets inside the human body, for a variety of diseases,

TABLE 5.2

List of Contrast Mechanisms Offered by the Different Imaging Modalities

Endogenous Mechanisms	Exogenous Mechanisms
Absorption, reflection, transmission (hard and soft photons)	Absorption, reflection, transmission, and emission (hard and soft photons)
Autofluorescence	Fluorescence
Polarization, refractive index	Bioluminescence
Magnetic relaxivity, magnetic susceptibility, magnetic spin tagging, proton density	Electroluminescence
Rayleigh scattering, Mie scattering, Raman scattering, Compton scattering	Chemoluminescence
Acoustic impedance	Hypoxia
Water molecule diffusion	Spin hyperpolarization
Oxygenation	Magnetic relaxivity, magnetic susceptibility
Spatial, spectra, temporal distribution	Magnetization transfer
Temperature	Saturation transfer
Electrical impedance	Isotope spectra, perfusion
Mechanical elasticity	Extracellular pH
Metabolites, biomarkers	Targeted probe technology

such as breast cancer, skin cancer, lung cancer, and bladder cancer; the study of drug effects on the target pathology and drug treatment effects; the development of biomarkers and molecular contrast agents indicative of disease and treatment outcomes; and the analysis of molecular pathways leading to diseases. Image formation through detection of the polarization states of light would offer distinct advantages for a wide range of detection and classification problems, given the intrinsic potential of optical polarimetry to offer high-contrast, high-specificity images under low-light conditions. On the clinical side, optical polarimetry would enhance imaging and spectral polarimetric information regarding the metabolic activities of the human body, as well as the molecular mechanisms, drug–cell interactions, single-molecule imaging, and so on. On the other side, a drawback of optical imaging is that light can only penetrate to a limited depth of tissue and also is not capable of providing high-resolution anatomic images with adequate spatial information of the target [46,49]. Traditionally, fluorescent imaging agents have been used in optical imaging. These imaging agents have some drawbacks such as quick photo bleaching, disability of simultaneous multiple targets detection, and cross-talk. Indeed, only a small number of fluorescent imaging agents can be deployed in NIR spectrum, and this is a limiting factor of using low-power energy lasers for interrogating specimens [12]. Today, imaging agents for optical imaging are called QDs, but disadvantages of optical imaging, which we discussed earlier, limit the usage of QDs. To overcome these weaknesses, improve the light penetration, and increase the spatial resolution, QDs can be used as imaging agents for MRI by coupling with paramagnetic material such as gadolinium [46,50]. The results show that contrast of in vitro images has been improved, but the potential ability of in vivo imaging of these nanoparticles is uncertain because of the lipid coating used to associate the gadolinium with the QD. The new type of QDs is promising for dual imaging. New studies on the biocompatibility of these nanoparticles gain a lot of interest. These types of QDs use manganese and are naturally magnetic and fluorescent that are suitable for dual imaging. QDs can conjugate simultaneously with different functional molecules such as antibodies and imaging probes. For example, in an experiment, the development of a new class of bioconjugated QD probes (appropriate for in vivo imaging) was reported. This new type of QDs used in this experiment were composed of amphiphilic tri-block copolymer used for in vivo protection, targeting-ligands in order to recognize tumor antigen and several PEG molecules to increase biocompatibility effect and improve circulation. The use of the bioconjugated QDs was reported for in vivo imaging of human prostate cancer cells that were growing in mice [51,52]. In another experiment, QDs were concatenated to translocation peptide in order to deliver them to cancer cells [53].

On the other hand, MRI is maturing to offer oncological images with improved soft tissue contrast and anatomical detail, while offering at the same time precious metabolic information. For instance, looking at the past, MRI was not an appropriate modality in the detection of small transient events such as lymph node metastasis, but today this imaging technique has progressed so that it offers anatomical detail of solid tumors and enhanced diagnostic images. Iron oxide is the most common metal used in the composition of nanoparticles as an imaging agent in MRI. Maghemite and magnetite are the two most common types of iron oxide. Between these two, magnetite is more biocompatible. These materials exhibit magnetic polarization in the presence of magnetic field. The magnetic properties of magnetite are so strong, and this feature makes it an appropriate contrast agent for MRI. In an experiment using iron oxide nanoparticles, even though these nanoparticles were unmodified, 90.5% detection of lymph node metastasis in a prostate cancer patient was reported by Harisinghani et al. while the conventional MRI was able to detect only 35.4% of metastasis nodes [4]. Superparamagnetic iron oxide NPs (SPIONs) are inorganic-based particles having either inorganic or organic coating materials. These NPs are used in clinical practice for hepatic, cellular, lymphatic, and cardiovascular imaging and have lower in vivo toxicity [12]. Liver lesions can be characterized using the SPIONs. Most healthy liver tissues contain RES [47,48,61]. SPIONs are accumulated by healthy cells while most liver tumors do not contain RES, so they are not taking SPIONs and the contrast between the healthy tissues, which have high intensity, and cancerous tissues, which are low-intensity sections of image, will be increased [12].

One of the significant factors in the battle against cancer is the early detection of cancer cells. Any specific molecular alteration of a cell on the DNA, RNA, metabolite, or protein level may be referred to as a molecular biomarker. According to the National Cancer Institute, biomarker is defined as "a biological molecule found in blood, other body fluids, or tissues that is a sign of a normal or abnormal process, or of a condition or disease." Alternatively, a cancer biomarker is a molecular signature that indicates the physiological and pathological changes in a particular tissue or cell type during the evolutionary stages of cancer. Cancer biomarkers can be used as precursors to identify the early onset of cancer as well as facilitate the diagnosis and/or patient management by providing quantitative figures aimed to accurately refine the staging of cancer, anticipating or monitoring response to treatment [112]. There is a critical need for expedited development of biomarkers to improve diagnosis and treatment of cancer.

From a practical point of view, the biomarker would specifically and sensitively map and quantify a disease state and could be used for diagnosis as well as for disease monitoring during and following therapy [6]. Biomarkers can be found in tissues and biofluids such as blood, urine, tumor cells, cerebral spinal fluid, and serum as well as in breath. Biomarkers can be classified as genetic biomarkers, proteomic biomarkers, metabolic biomarkers, and breath biomarkers [118]. The biomarkers may be genes, proteins, metabolites, and breathing ingredients, which provide information of abnormal growth of cells in the cancerous cells. Various detection techniques rely on using analytical chemistry techniques, such as chromatography, enzyme-linked immunosorbent assay (ELISA), two-dimensional gel electrophoresis (2DE), immunohistochemistry (IHC), fluorescence in situ hybridization (FISH), polymerase chain reaction (PCR), real-time polymerase chain reaction (RT-PCR), matrix-associated laser desorption/ionization time-of-flight MS (MALDI-TOF-MS), surface-enhanced laser desorption/ionization time-of-flight MS (SELDI-TOF-MS), and liquid chromatography coupled with various interfaces, such as LC/MS, LC/MS/MS, and others [118,119].

Interestingly enough, the development of LC/MS significantly impacted biological research, including metabolomics, due to its ease of operation, inexpensiveness, selectiveness, reproducibility, and low limits of detection [117]. Originally, gas chromatography was the only separation method able to be hyphenated to mass spectrometry; however, it is limited only to biological molecules of low molecular weight that are volatile; alternatively, they should be derivatized, excluding, therefore, biological molecules of high molecular weight, such as proteins or nucleic acids. The introduction of atmospheric pressure ionization mass spectrometry (API-MS)–based techniques combined with liquid chromatography, which exhibit a good sensitivity, high dynamic range, and versatility but also provide soft ionization conditions giving access to the molecular mass of intact biological molecules, leading to a significant enhancement of the biomolecule detection. One of the strengths of API-MS-derived tool is the high diversity of analyzers available: triple quadrupoles, ion traps, time of flight, orbitrap, and Fourier transform-ion cyclotron resonance instruments, the latter three providing high-resolution and accurate mass measurements [118,119]. High-resolution analyzers are becoming increasingly popular in the field of metabolomics because they provide (a) accurate mass measurement and (b) structural information with MS/MS, or sequential MSn experiment, especially when ion products are analyzed at high resolution [43]. A diagram offering an overview of mass spectrometers and separation techniques accompanied by their respective resolving power and separation power is shown in Figure 5.11.

On the other hand, liquid chromatography-tandem mass spectrometry (LC-MS/MS) provides excellent analytical specificity superior to that of immunoassays or conventional high-performance/pressure liquid chromatography (HPLC). It is suitable for low-molecular-weight analytes and has a higher throughput than gas chromatography-mass spectrometry (GC-MS) [117,119].

Moreover, cancer biomarker discovery has greatly benefited from the proteomics [112–116]. The technological evolution from the 2DE into the liquid chromatography (LC)-based high-resolution tandem mass spectrometry (MS/MS) has introduced a significant improvement on the speed and precision of identifying and quantifying target proteins in biological fluids and other samples. Proteomic research relies on the synergistic efforts of several analytical. A careful analysis of biomarkers such as protein is extremely important in the diagnosis, management, and monitoring of cancer. Recently, the development of nanowire (NW) provides the possibility of detecting the cancer in early stages when the cancer cells are a few thousands. They have the ability to detect few proteins along with other biomarkers [1]. Specifically, Silicon nanowires (SiNWs) showed that this platform is promising in the area of label-free and real-time electrical detection of proteins and other biomarkers with high sensitivity [40,43].

The following is the list of few tumor biomarkers [39]:

Prostate cancer: PSA (prostate-specific antigen)

Ovarian cancer: CA 125 (elevated cancer antigen), HCG, CEA, p53, CASA, CA19-9

Breast cancer: CA 15-3, CA 125, Her2/NEU

Lung cancer: CEA, CA 19-9, SCC, NSE, NY-ESO-I

Overall, the proteomics cancer research including applied analytical protocols is depicted in Figure 5.12.

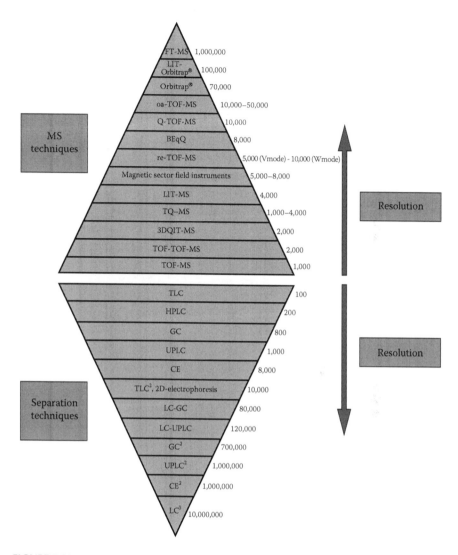

FIGURE 5.11

An overview of mass spectrometers and separation techniques accompanied by their respective resolving power and separation power.

5.4.1 QD-Based Detection of Primary Tumor

Ovarian cancer is the second common cancer in females. In experiment by Wang et al., QDs with a maximum emission wavelength of 605 nm were used to detect carbohydrate antigen 125 (CA 125), which is an epithelial antigen and is useful to label the tumor in the process of detection and

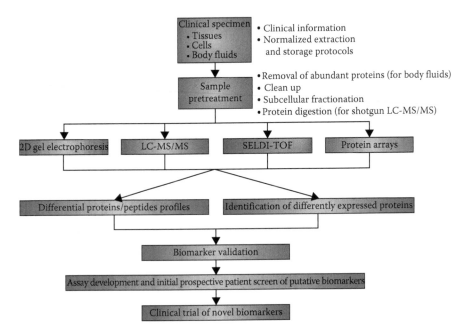

FIGURE 5.12

Proteomics cancer research and applied procedures. (From Ge, F. and He, Q.-Y., Proteomics in cancer biomarker discovery, in: X. Zhang, ed., *Omics Technologies in Cancer Biomarker Discovery*, Landes Bioscience, Austin, TX, 2011, Chapter 3, pp. 39–48.)

therapy of ovarian cancer [25–27]. The results of experiments show that signals of QDs labeling had higher brightness in comparison with fluorescein isothiocyanate (FITC). These signals were also more stable and specific. In another study, biocompatible QDs coated with silk fibroin (SF), which is a natural protein, were used for bioimaging HEYA8 ovarian cancer cells [28]. These QDs behave as fluorescent label and are more luminescent and stable.

Breast cancer QDs were used for immunofluorescent labeling of HER2 (human epidermal growth factor receptor 2). Wu et al. synthesized the QDs with a polyacrylate and covalently connected to antibodies. Compared with other fluorescent markers, QD signals were more stable and brighter and were more specific. Researches on QDs have led us to the development of bioconjugate QDs that are remarkably luminescent [29]. Several other studies have been done for breast cancer diagnosis with the aim of detecting HER2 using QDs [30,31]. Yezhelyev et al. used multicolor QDs to detect important tumor markers in breast cancer cells MCF-7 and clinical specimens BT474 [24,32]. Five significant tumor markers were simultaneously profiled and detected in these set of experiments: HER2 (QD-HER2), ER (QD-ER), PR (QD-PR), EGFR (QD-EGFR), and mTOR (QD-mTOR).

Lymph node metastasis investigation is a significant factor in cancer prognostic, and state of tumor draining lymph node is so important for the diagnosis of cancer. Among the lymph nodes of the same scope, sentinel lymph node (SLN) has the higher risk of containing tumor cells [33,34]. There are several methods for the diagnosis of SLN such as QDs, which received much interest as the delivery agents of lymph node [35]. Near-infrared (NIR) QDs are useful for the diagnosis of lymph nodes metastasis because they have the high ability of tissue penetration and low background [36,37]. Ballou et al. injected QDs into two model tumors in an experiment and showed that QDs were passed to sentinel lymph nodes quickly [38]. Drifting of QDs to the lymphatic system was observed by imaging during necropsy. This approach also provided useful information for pathology examinations by tagging the SLNs. Other advantage of examining the SLNs using QDs was the observation of metastatic tumor foci in some nodes [24].

5.5 Paradigm: Nanoscale Flow Imaging

5.5.1 Physical Principles

Contrast measurements become of increasing importance in digital imaging, where the region of interest (ROI) differences can be effectively identified, processed, and segmented. The image contrast among different structures varies with the material properties, physical, chemical, electrical, optical parameters, and it is difficult to be determined only from its composition and geometrical parameters.

Moreover, optical polarimetry relies on the properties of polarization of backscattered light and results in distinct signatures related to surface smoothness, orientation, and sample composition, physical, chemical, and optical characteristics.

Based on these considerations, a novel molecular-based optical imaging technique has been proposed by Giakos [82–88]. Specifically, he proposed that further contrast enhancement of the target can be achieved by modulating the background of the biological structure through doping with nanoparticles, polar and high-index-of-refraction molecules. Utilizing the rotation of the polarization vector enforced by optically active (chiral) molecules, together with efficient polarimetric interrogation techniques, yields both optical clearing and enhanced contrast capabilities, as initially reported in [85–88]. This technique relies on the synergistic efforts of doping the background surrounding the biological structure with optical active molecules or high-index-of-refraction molecules, in conjunction with the application of efficient polarimetric interrogation techniques. As a result, the proposed technique would provide both optical clearing and enhanced contrast capabilities; minimizing refractive index differences between the optically active fluid medium and the target would result in an increase of the degree of polarization (DOP), with an increase in the concentration of the optically active molecules, at the expense of the contrast resolution. On the other hand, enhanced image contrast would result using Stokes parameter imaging because of its intrinsic potential to detect weakly backscattered linearly polarized radiation in the presence of highly backscattered depolarized radiation, while maintaining the original state of polarization of the incident light. Overall, using high-refractive-index dielectric particles dispersed within fluids would cause the refractive index of the liquid to become a volume-weighted average between refractive indices of optically active molecules and the liquid phase, ultimately giving rise to an amplification of the index of refraction, yielding enhanced DOP and degree of linear polarization (DOLP) signal-to-background ratio (SBR) [85–101].

In this study, the outcomes of two different experiments reported in [87,88] were presented and analyzed, where the reference images include a structure (target or phantom) immersed into aqueous solutions surrounded by optically active nanomolecules (dopants). In order to assess the potential of the applied methodology, the DOP was estimated by measuring the Stokes parameters of the backscattered light by means of the *Fourier Analysis using a Rotating Quarter-Wave Retarder Method* [86,87]. The novelty of this study consists in fusing statistical analysis with polarimetric principles. As a result, quantification of image contrast in terms of Stokes parameters together with the modeling of intensity distribution for the corresponding target areas can be proved to be a powerful tool for analyzing the different properties of operational modalities and/or materials depicted in digital images. By fusing these concepts, we explored the intrinsic potential of an efficient nanofluidics-based imaging technique aimed at increasing the optical contrast of a structure surrounded by a scattering medium.

5.5.2 Polarimetric Fusion with Statistical Models

5.5.2.1 Models and Applications

Following the treatment of [82,87] by Giakos, the polarization interaction with the scattering medium, displayed as a transformation of the polarization state, is captured by a CCD camera, producing the polarimetric images. The state of polarization of a target can be characterized by its four Stokes parameters (S_0, S_1, S_2, S_3). The DOLP can be estimated in terms of Stokes parameters (S_0, S_1, S_2) as

$$\text{DOLP} = \frac{\sqrt{S_1^2 + S_2^2}}{S_0} \qquad (5.11)$$

while the DOP is given as

$$DOP = \frac{\sqrt{S_1^2 + S_2^2 + S_3^2}}{S_0} \qquad (5.12)$$

One of the key signatures for material characterization is the depolarization of a completely polarized state with DOP equal to 1, expressed as

$$1 - DOP(\mathbf{S'}) = 1 - DOP(\mathbf{MS}) \qquad (5.13)$$

where
  $\mathbf{M}$ is the Mueller matrix of the object
  $\mathbf{S'}$ and $\mathbf{S}$ are the Stokes vectors of the detected and impinging to the object Stokes vectors

The angle φ of the rotation of the polarized light, when passed through optically active molecules, depends on their number in the solution [92,93]; thus, it depends on both the concentration and the path length through the substance.

The novelty of the proposed methodology combining polarimetry with digital image enhancement is twofold [87]: (1) it can quantify the efficiency of polarimetric imaging modalities in discriminating materials of interest and, (2) it can enable improved segmentation and discrimination of materials through the enhancement of fitted-model differences. Specifically, a thorough analysis of the digital image distribution is performed based on fitting the intensity histogram of the acquired images using mixture models [89], revealing additional characteristics of the polarized object structures. Overall, the proposed methodology is applied on different histogram forms of DOLP images on the three sets of experiments. The physical purpose of these experiments is examining the effect of different chiral or high-index-of-refraction nanomolecules (glucose and L-phenylalanine) when used as molecular contrast agents, in the image quality of a detected target immersed in an aqueous solution. The tangible outcome of our study is that enhanced DOLP is achieved with increasing the concentration of the optically active molecules, providing both optical clearing and enhanced contrast capabilities.

The model description of a histogram from a particular object can be used to effectively assign local pixels to the region of that object through the corresponding probability implied by the fitted statistical distribution. Normally, there are a variety of possible distributions that may serve as basic functions for the fitted model, which can be evaluated through theoretical or computational means [94–96]. In order to determine how well the estimated model approximates the initial distribution, several criteria have been successfully used for revealing the *goodness of fit*. As such a metric for our model we use the mean squared error (MSE), quantifying the amount by which the estimated curve $\bar{X}$ differs from the original one X, as

$$E = \frac{1}{n-1}\sum_{i=1}^{n}\left(X_i - \bar{X}_i\right)^2 \tag{5.14}$$

where n is the number of samples i. We also consider the dB gain in the signal-to-noise-ratio (SNR), defined as

$$SNR = 10\log_{10}\frac{\sum_{i=1}^{n}X_i^2}{\sum_{i=1}^{n}\left(X_i - \bar{X}_i\right)^2} \tag{5.15}$$

Finally, as figure of merits (FOMs) aimed at verifying the detection rates of our experimental result, we adopted the accuracy, specificity, and sensitivity of the object discrimination procedure. Interestingly enough, accuracy defines the proportion of true results in the classification procedure, sensitivity measures the proportion of actual positives that are correctly identified as such, and specificity represents the proportion of negatives that are correctly identified.

5.5.2.2 Computational Algorithm for Single- and Multimodal Distributions

Depending on the properties of the imaging modality, the regions of different materials in the recorded image can be modeled and discriminated by means of the intensity or the texture of the corresponding region. For instance, two materials that absorb different portions of the incident radiation would be recorded as image regions of quite different intensities. In this case, the use of mixture models applied directly on the intensity of a smoothed intensity histogram could enable the separation of two distinct models, one for each material region. Alternatively, if the imaging modality favors the dispersion and extensive diffuse of light at grain levels, then the texture of different material regions would be preferred as a discriminant factor. In such cases, the modeling of the distribution of variance as an expression of texture would be most appropriate for characterizing the structure of material regions [97].

In this study [87,88], we consider the outcomes of two experiments that appropriate the modeling of either intensity or texture characteristics. In order to suppress possible noise contamination in the intensity distribution, we process the local mean images obtained from 5×5 moving average filtering. In a similar way, the local variance images are acquired through the application of a 5×5 moving variance window shifted along the pixels of the entire image. The discrimination of image regions based on mean-intensity differences suggests the use of Gaussian-like distributions, while the discrimination based on texture measures such as variance necessitates the use of the chi-square distribution according to the theoretical analysis of variance. In essence, the Gaussian model for mean-intensity distribution is justified by either the assumption of a compact region for each material, as in [94], or the central-limit theorem guiding the distribution of the local mean operator [95]. Furthermore, the Gaussian assumption by itself enforces the chi-square distribution on the local variance estimates [98].

Using the Gaussian and the chi-square distributions as basis functions, we can define the mixture of Gaussian and the mixture of chi-square distributions for the mean intensity and variance, respectively [88]. The input images to be processed include a structure of interest (or target) and a surrounding region (or background). Due to the camera lens, the recorded image involves a circular ROI and a supplementary area at the corners of the rectangular image, which is an irrelevant area. Thus, the mixture of Gaussians can be represented as a sum of three distributions: one that represents the target, one that represents its surrounding medium, and one that represents the remaining region of noninterest, according to the following form:

$$f(x; A_i, \mu_i, \sigma_i) = \underbrace{A_1 \cdot \exp\left(-\frac{(x - \mu_1)^2}{2\sigma_1^2}\right)}_{\text{Target}} + \underbrace{A_2 \cdot \exp\left(-\frac{(x - \mu_2)^2}{2\sigma_2^2}\right)}_{\text{Background}}$$

$$+ \underbrace{A_3 \cdot \exp\left(-\frac{(x - \mu_3)^2}{2\sigma_3^2}\right)}_{\text{Remaining area}}, \tag{5.16}$$

where

A_1, A_2, A_3 are the amplitude factors

μ_1, μ_2, μ_3 are the means

$\sigma_1^2, \sigma_2^2, \sigma_3^2$ are the variances of the distributions of the target, surrounding medium, and remaining area, respectively

If we model the circular ROI only, then the last function in the model can be eliminated. Similarly, the mixture of (two) chi-square functions of interest can be represented as

$$f(\chi^2; A, n) = \underbrace{A_1 \cdot \frac{1}{2^{(n_1/2)} \cdot \Gamma(n_1/2)} \cdot (\chi^2)^{(n_1/2)-1} \cdot e^{-\chi^2/2}}_{\text{Target}}$$

$$+ \underbrace{A_2 \cdot \frac{1}{2^{n_2/2} \cdot \Gamma(n_2/2)} \cdot (\chi^2)^{(n_2/2)-1} \cdot e^{-\chi^2/2}}_{\text{Background}}, \tag{5.17}$$

where

  $\Gamma(\cdot)$ is the gamma function

  A_1, A_2 are the amplitude factors

  n_1, n_2 are the degrees of freedom of the distributions of the target and the
    surrounding medium, respectively

The term *degree of freedom* is defined as the number of terms in the final calculation of a statistical problem, herein the chi-square distribution, that we can vary freely. The distribution of the local variance image of the remaining area has no significant contribution as this region appears homogenous; thus, its variance equals to values near zero. The mean of the χ^2 distribution is n, its variance is $2n$, and its mode equals to $n - 2$ [98]. As the degrees of freedom increase, the chi-square function approximates the Gaussian one.

  Having accurately defined the model of our hypothesis, the estimation of its optimal parameters is then performed in the least-square sense. Thus, we need to find parameters vector $\mathbf{p}$ that best fits the equation (minimize the least square error):

$$\min_{\mathbf{p}} \frac{1}{2} \| f(x;\mathbf{p}) - y(x) \|_2^2 = \min_{\mathbf{p}} \frac{1}{2} \sum_{i=1}^{m} \left(f(x_i;\mathbf{p}) - y(x_i) \right)^2 \tag{5.18}$$

where

  x is the input data vector of size m (in our case $m = 255$, which is the number
    of the different gray intensity levels)

  y is the observed histogram of x

  $f(\cdot)$ is the hypothetical model function

  $\mathbf{p}$ is the parameters vector to be optimized

When we perform comparisons of different material types or concentrations under the same experimental setup, then we might need to scale the mixture models under comparison so that the background mean remains fixed. In such cases, a scaling procedure is also implemented, which aligns the background means but otherwise preserves the shapes of model distributions.

5.6 Contrast Discrimination Measures

The acquired contrast is an attribute of the imaging modality used and can be quantified by metrics applied on the recorded image [88]. The better the contrast achieved, the better the effectiveness in correctly segmenting the regions associated with different materials in the image. Furthermore, histogram processing of the digital image can further contribute to contrast enhancement. In order to quantify robust contrast metrics, we employed the statistical modeling schemes of the polarimetric input images. In the previous section, we modeled the distribution of intensities of the various materials depicted in the acquired image. Here, we associate the image contrast with the separability of model distributions. The distance of the distribution centers reflects the structural differences of the two regions (target and surrounding medium). Thus, we propose to use the difference of modes (DoM) as a contrast measure of images. The larger the difference, the better the two regions are discriminated, which reveals contrast enhancement. For the mixture of Gaussian model, the contrast metrics is defined as the difference of the distribution means (equivalent to modes). Alternatively, for the mixture of chi-square functions, the contrast measure is computed as the difference of the modes of the foreground and the background distributions [88].

5.6.1 First Set of Experiments (L-Phenylalanine in Aqueous Solution at 830 nm)

In order to compare the effects of the optically active molecules at different concentration levels, we model the distributions of the acquired images. The experimental setup and the applied procedures are adequately described in [87]. Within the area of interest, marked by a circle, we find low-intensity water segments mixed with the optical dopant at higher-intensity levels. Due to the high magnification of the electron microscope, however, these regions are heavily contaminated by noise, which masks the intensity distributions of individual materials. In order to partially alleviate this problem, we process the local mean images obtained from 5×5 moving average filtering. This filtering fulfils the required condition analyzed in [95] about the smoothing of the gray-level histogram in order to perform distribution modeling via Gaussian-like distributions. The resulting image is depicted in the upper middle part of Figure 5.13.

The corresponding histogram in the lower middle part of Figure 5.13 can be approximated as a mixture of three Gaussian distributions: one distribution representing the target/foreground (the plastic twisted wire), one distribution representing the surrounding medium/background (L-phenylalanine molecules in the glass test tube containing water), and one distribution for the remaining area outside the circle, according to Equation 5.9. This modeling scheme using the mixture of Gaussians is essential under the assumption of Gaussian noise contamination in the original image. The distance of distribution means reflects the structural differences of the two regions. Thus, we use the difference of model means as a contrast measure of images at different concentrations.

Furthermore, we also consider the local variance image (upper-right part of Figure 5.13) in an attempt to exploit the texture differences between the (smooth) aqueous solution and (more active) the twisted polystyrene phantom area. For the variance image, the appropriate distributions characterizing the two regions become chi-square (χ^2) distributions. From both forms

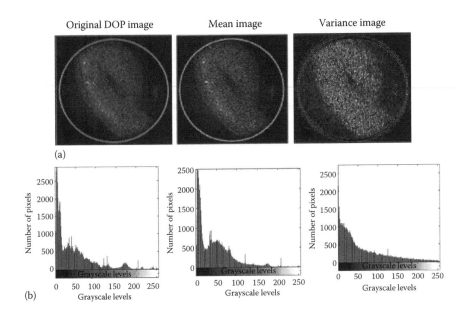

FIGURE 5.13

(a) Original, mean, and variance images along with (b) their corresponding histogram distributions for a certain concentration of phenylalanine in aqueous solution.

of the image and its histogram in Figure 5.13, we can verify that the distribution of the background (surrounding medium) imposes fewer degrees of freedom (smaller variance values) than the distribution of the foreground (target), which is spread over the upper part of the dynamic range, at high values. We model the entire image distribution through a mixture of two χ^2 functions estimated by best fit on the histogram through least-squares optimization and use their modes in the proposed contrast measure. The difference of model modes in the variance image is used as an additional contrast measure at different concentrations.

The two contrast measures for the different concentrations of phenylalanine are depicted in Figure 5.14, indicating a clear contrast increase with concentration. The difference of modes for either the mean or the variance images has been considered with respect to the scaled range of [0, 255].

The original and the model-estimated histograms are shown in Figure 5.15a for the local mean and variance images at the left and right parts, respectively.

The proximity of the two curves, as well as the effectiveness of our approximation, is supported by the *goodness-of-fit metrics*, reflecting an MSE value of order 5×10^{-3} and an SNR value on the order of 10 dB. The estimation procedure using the mean image achieved an average value for accuracy equal to 97.03% for the six distinct concentrations used. Furthermore, the measures of specificity and sensitivity are on the order of 88.21% and 99.54% respectively,

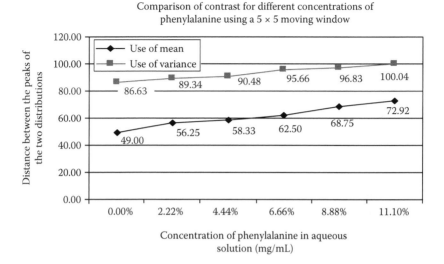

FIGURE 5.14
DoM for the mean and variance images versus concentration of aqueous L-phenylalanine. (From Giakos, G. et al., *Meas. Sci. Technol.*, 20, 12, 2009.)

clearly demonstrating the capability of effectively detecting both regions of interest, with a prevalent accuracy of the target one. By decomposing the estimated mixture models, we can easily derive the image regions being modeled by each individual distribution. The corresponding target and background regions using the mean and variance signals are illustrated in the second and third rows of Figure 5.15, respectively.

The mean image appears more robust in segmenting the target regions of interest, whereas the variance image segments better the surrounding medium regions. Thus, the two forms of complementary images can be used in order to best segment all regions of interest. At this point, we should notice that by further modifying the fitted models we can obtain enhanced separation of materials and better discrimination of the regions of interest.

5.6.2 Second Set of Experiments (Glucose in Aqueous Solution at 655 nm)

To perform the experiment with increasing the glucose dopant, an aqueous glucose solution was prepared at first by adding 12.5 g of glucose to 100 mL of water. The experimental setup and applied procedures are reported in [87]. The glucose solution was then added to the phantom that contained 16.8 mL of water in concentration increments of 10.42 up to 52.08 mg/mL. For each concentration increment of glucose solution, the co-polarized and cross-polarized images along with the DOLP images were obtained. An example of such images is illustrated in the upper-left part of Figure 5.16. Within the

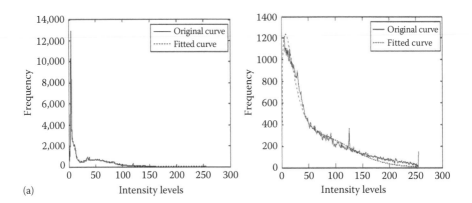

Original and fitted histograms of the mean (left) and variance (right) images

(a)

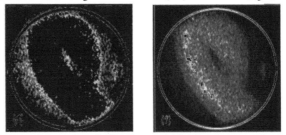

Surrounding medium (left) and target (right) based on mean image

(b)

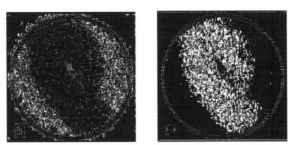

Surrounding medium (left) and target (right) based on variance image

(c)

FIGURE 5.15

(a) Original and fitted curves for mean and variance images, along with discriminated image regions for the (b) mean and (c) variance images. (From Livanos, G. et al., *IET Image Process.*, 5(5), 429, 2011.)

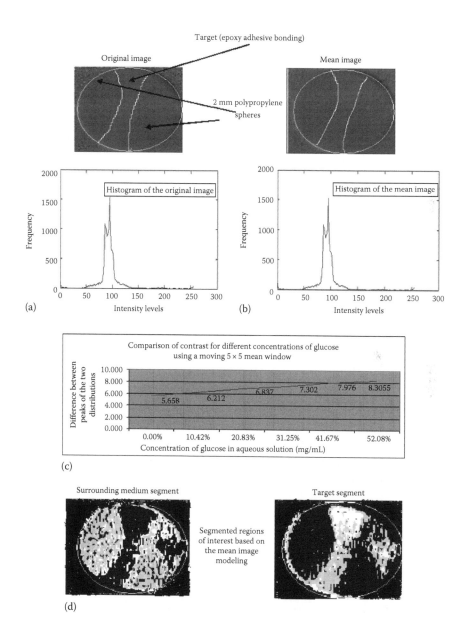

FIGURE 5.16
Original and mean processed DOLP images along with their corresponding histogram distributions (a and b, respectively); DoM contrast measure with increasing concentrations of aqueous glucose (c); segmented regions of interest for surrounding medium and target (d). (From Giakos, G. et al., *Meas. Sci. Technol.*, 20, 12, 2009.)

circular area of focus, two sectors of index of refraction $n = 1.49$ are depicted, separated by a strip of index of refraction $= 1.65$.

Attempting to quantify the contrast effects with increased concentration, we consider the proposed statistical modeling of the DOLP images. From the appearance of the images at this wavelength and magnification level, it becomes obvious that the information at various segments can only be differentiated in terms of intensity levels. Within the circular area of focus, the texture of the various subregions is smooth and similar everywhere, thus rendering the local variance quite inefficient for region differentiation. Instead, the illumination levels of the regions of interest are different due to the different absorption characteristics, appropriating the use of mean intensity as an efficient metric for discrimination.

Toward this direction, we processed the local mean images obtained from 5×5 moving average filtering (upper-right part of Figure 5.16). The corresponding histogram distributions are presented in Figure 5.16b. The averaging operator smoothens out the intensities of regions, making their histograms more compact and appropriate for modeling, as the conditions for the adoption of mixture of Gaussian-like distributions are fulfilled [95]. The local mean image reveals two distinct, rather symmetric distributions that can be efficiently modeled by normal distributions, representing the target and its surrounding medium, while a third one models the remaining area outside the circle of interest. Therefore, the entire histogram distribution can be modeled as the sum of three Gaussian functions. The distance of distribution means reflects the structural differences of the two regions. We use the difference of model modes (means) as a contrast measure of images at different concentrations. The results for increasing the glucose concentration in aqueous solution are illustrated in Figure 5.16c, clearly quantifying the contrast improvement over increased concentration levels. Since the contrast of the acquired images is not high, we also illustrate the efficiency of the modeling process to correctly quantify the regions of interest. The segmented areas of the polypropylene spheres and the adhesive bonding structure are depicted in Figure 5.16d. Notice that in this experiment the contrast of material structure is low, even for high concentration of glucose. This is clearly illustrated in the comparative DoM plot at the third row of Figure 5.16. The goodness-of-fit metrics reflects an MSE of 2×10^{-3} and an SNR dB gain value of order 10. The classification metrics for the six distinct concentrations of glucose revealed an average value for accuracy, specificity, and sensitivity of 93.66%, 92.23%, and 98.87%, respectively. The relatively reduced accuracy in this case is due to the complex 3D structure of the scene (polypropylene spheres with adhesive bonding target) that affects the illumination distribution on the spheres and causes false detection of some background as target region.

In summary, intrinsic qualities of biological media, the surrounding fluids, have an effect on the way light interacts with them, whether in transmission or reflection. Polarimetric imaging can provide complementary discriminative information in applications such as object detection. The contrast of different materials, however, is difficult to be quantified and is used in further comparative studies. In summary, in this study, an efficient tool for contrast quantification by means of histogram modeling has been introduced. More specifically, first the histogram of a recorded DOLP image as a mixture of statistical functions is set, each representing a specific region within the original scene. Using the parameters of such models, a contrast metric based on the difference of modes of the fitted distributions was defined. Based on the proposed modeling scheme, a novel imaging technique is explored aimed to provide both improved penetration depth and imaging contrast. The outcome of this analysis indicates that enhanced image contrast capabilities result in increasing the concentration of the high index of refraction/optically active molecules.

5.6.3 Therapeutics

In PDT, light absorption of a light-activatable chemical known as a photosensitizer results in the generation of cytotoxic oxygen-based reactive species that can cause cellular damage and cell death via oxidative stress. The effectiveness of PDT is determined by singlet oxygen production and tumor-cell-selective targeted delivery of photosensitizers. PDT has recently applied for several cancers, including skin [62], bladder [63], prostate [64], lung [65], esophageal [66], pancreatic [67], stomach [68], and head and neck [69] cancers.

Photothermal therapy refers to hyperthermic treatment of tumors. Specifically, temperatures above 42°C can induce irreversible tumor destruction, causing apoptosis and cell death by loosening membranes and denaturing proteins. NIR, laser, RF, microwaves, magnetic fields, and ultrasound are utilized for controlled and selective heating of nanoparticles such as gold nanoshells combined with targeting proteins that are delivered to the tumor site for targeted thermal damage to the tumor with minimal damage to surrounding normal tissues. The effectiveness of photothermal therapy depends on light absorption and light-to-heat conversions of the nanoparticles. Applications of photothermal therapy have been reported in prostate cancer [107], malignant glioma [108], and breast cancer [109], and most recently brain cancer in the use of magnetic fluid hyperthermia that consists of iron oxide nanoparticles within water. Nanobody conjugated to enzymes such as *Enterobater cloacae* beta-lactamase is capable of site-selective anticancer prodrug activation. Nanobombs are nanoscale bombs that can be exploded under light exposure. The explosion of nanobombs creates shockwaves and heat that can kill cancer cells and then are effectively cleared by macrophages. Both nanobody and nanobombs are minimally invasive and potential approaches for cancer therapeutics [8–10,15,107–109].

5.7 Conclusion

In this chapter, the principles of molecular imaging and omics imaging and spectroscopy techniques on the cancer detection were introduced. Fluid dynamics modeling of the tumor vasculature and drug transport were discussed and analyzed. The properties of nanoscale particles and their impact on the diagnosis, therapeutics, and theranostics were introduced and analyzed. Combining nanotechnology and nanoscience principles together with molecular imaging, and omics disciplines, nanooncology is poised to revolutionize the clinical arena by introducing efficient, and reliable methodologies for the diagnosis, assessment, monitoring, treatment, and management of cancer.

References

1. F. Leonard, Imaging and cancer: A review, *Mol Oncol* 2, 115–152, 2008.
2. X. Sun, Z. Liu, K. Welsher, J. T. Robinson, A. Goodwin, S. Zaric, and H. Dai, Nano-graphene oxide for cellular imaging and drug delivery, *Nano Res* 1, 203–212, 2008, doi: 10.1007/s12274-008-8021-8.
3. M. Hofmann-Amtenbrink, B. von Rechenberg, and H. Hofmann, Super paramagnetic nanoparticles for biomedical applications, M. C. Tan (ed.), *Nanostructured Materials for Biomedical Applications* 2009: Transworld Research Network, 120–148, 2009, ISBN: 978-81-7895-397-7.
4. L. E. van Vlerken and M. M. Amiji, Multi-functional polymeric nanoparticles for tumour targeted drug delivery, *Expert Opin Drug Deliv* 3(2), 205–216, March 2006, doi:10.1517/17425247.3.2.205.
5. F. Ge and Q.-Y. He, Proteomics in cancer biomarker discovery, X. Zhang (ed.), *Metabonomics in Cancer Biomarker Discovery*, Landes Bioscience, Austin, TX, 2011, pp. 49–51, Chapter 4.
6. P. Koumoutsakos, I. Pivkin, and F. Milde, The fluid mechanics of cancer and its therapy, *Annu Rev Fluid Mech* 45, 325–355, 2013, doi: 10.1146/annual review fluid 120710-101102.
7. H. Shimizu, R. Ishibashi, K. Mawatari, and T. Kitamori, Attoliter liquid chromatography using extended-nano channel for separation of proteins in a single cell, *16th International Conference on Miniaturized Systems for Chemistry and Life Sciences*, October 28–November 1, 2012, Okinawa, Japan, 2012.
8. L. Zhang, F. X. Gu, J. M. Chan, A. Z. Wang, R. S. Langer, and O. C. Farokhzad, Nanoparticles in medicine: Therapeutic applications and developments, *Clin Pharmacol Ther* 83(5), 761–769, 2008.
9. M. Ferrari, Cancer nanotechnology: Opportunities and challenges, *Nat Rev Cancer* 5(3), 161–171, 2005, doi: 10.1038/nrc1566.

10. Advances in the field of nanooncology—Nanotechnology against cancer, 2011 brian wang, Advances in the field of nanooncology—Nanotechnology against cancer, 2011, BMC Med. 2010;v 8: 83. Published online December 13, 2010. doi: 10.1186/1741-7015-8-83, http://nextbigfuture.com/2011/01/advances-in-field-of-nanooncology.html.

11. C. Riggio, E. Pagni, V. Raffa, and A. Cuschieri, Nano-oncology: Clinical application for cancer therapy and future perspectives, *J Nanomater* 2011, Article ID 164506, 10 pp., 2011.

12. A. S. Thakor and S. S. Gambhir, Nanooncology: The future of cancer diagnosis and therapy, *CA Cancer J Clin* 63(6), 395–418, 2013, doi: 10.3322/caac.21199.

13. S. Bhaskar, F. Tian, T. Stoeger, W. Kreyling, J. M. de la Fuente, V. Grazú, P. Borm, G. Estrada, V. Ntziachristos, and D. Razansky, Multifunctional nanocarriers for diagnostics, drug delivery and targeted treatment across blood-brain barrier: Perspectives on tracking and neuroimaging, *Part Fibre Toxicol* 7, 3, 2010.

14. S. Suresh, Nanomedicine: Elastic clues in cancer detection, *Nat Nanotechnol* 2, 748–749, 2007.

15. J. Xie, S. Lee, and X. Chen, Nanoparticle-based theranostic agents, *Adv Drug Deliv Rev* 62(11), 1064–1079, 2010, doi: 10.1016/j.addr.2010.07.009.

16. G. Giakos, Future directions: Opportunities and challenges, S. M. Musa (ed.), *Computational Nanotechnology: Modeling and Applications with MATLAB*, CRC Press, Boca Raton, FL, 2012.

17. B. Sumer and J. Gao, Theranostic nanomedicine for cancer, Future Medicine Ltd, *Nanomedicine*, 3(2), 137–140, April 2008, doi: 10.2217/17435889.3.2.137.

18. S. Chinnathambi, S. Chen, S. Ganesan, and N. Hanagata, Silicon quantum dots for biological applications, *Adv Healthcare Mater* 3, 10–29, 2014, doi: 10.1002/adhm.201300157.

19. M. N. Rhyner, A. M. Smith, X. Gao, H. Mao, L. Yang, and S. Nie, Quantum dots and multifunctional nanoparticles: New contrast agents for tumor imaging, Future Medicine Ltd, *Nanomedicine*, 1(2), 209–217, August 2006, doi: 10.2217/17435889.1.2.209.

20. R. D. Tilley and K. Yamamoto, The microemulsion synthesis of hydrophobic and hydrophilic silicon nanocrystals, *Adv Mater* 18, 2053–2056, 2006.

21. J. H. Ahire, Q. Wang, P. R. Coxon, G. Malhotra, R. Brydson, R. Chen, and Y. Chao, Highly luminescent and nontoxic amine-capped nanoparticles from porous silicon: Synthesis and their use in biomedical imaging, *ACS Appl Mater Interf* 4(6), 3285–3292, 2012, doi: 10.1021/am300642m.

22. F. Erogbogbo, K. T. Yong, I. Roy et al., In vivo targeted cancer imaging, sentinel lymph node mapping and multi-channel imaging with biocompatible silicon nanocrystals, *ACS Nano* 5, 413–423, 2011.

23. H. Zhang, D. Yee, and C. Wang, Quantum dots for cancer diagnosis and therapy: Biological and clinical perspectives, *Nanomedicine* 3(1), 83–91, 2008.

24. C.-W. Peng and Y. Li, Application of quantum dots-based biotechnology in cancer diagnosis: Current status and future perspectives, 2010, Article ID 676839, 11 pp., doi: 10.1155/2010/676839.

25. E. V. S. Høgdall, L. Christensen, S. K. Kjaer et al., CA125 expression pattern, prognosis and correlation with serum CA125 in ovarian tumor patients. From The Danish "MALOVA" Ovarian Cancer Study, *Gynecol Oncol* 104(3), 508–515, 2007.

26. R. W. Tothill, A. V. Tinker, J. George et al., Novel molecular subtypes of serous and endometrioid ovarian cancer linked to clinical outcome, *Clin Cancer Res* 14(16), 5198–5208, 2008.
27. Y. Zheng, D. Katsaros, S. J. C. Shan et al., A multiparametric panel for ovarian cancer diagnosis, prognosis, and response to chemotherapy, *Clin Cancer Res* 13(23), 6984–6992, 2007.
28. B. B. Nathwani, M. Jaffari, A. R. Juriani, A. B. Mathur, and K. E. Meissner, Fabrication and characterization of silk-fibroin-coated quantum dots, *IEEE Trans Nanobiosci* 8(1), 72–77, 2009.
29. X. Wu, H. Liu, J. Liu et al., Immunofluorescent labeling of cancer marker Her2 and other cellular targets with semiconductor quantum dots, *Nat Biotechnol* 21(1), 41–46, 2003.
30. S. Li-Shishido, T. M. Watanabe, H. Tada, H. Higuchi, and N. Ohuchi, Reduction in nonfluorescence state of quantum dots on an immunofluorescence staining, *Biochem Biophys Res Commun* 351(1), 7–3, 2006.
31. Y. Xiao, X. Gao, G. Gannot et al., Quantitation of HER2 and telomerase biomarkers in solid tumors with IgY antibodies and nanocrystal detection, *Int J Cancer* 122(10), 2178–2186, 2008.
32. M. V. Yezhelyev, A. Al-Hajj, C. Morris et al., In situ molecular profiling of breast cancer biomarkers with multicolor quantum dots, *Adv Mater* 19(20), 3146–3151, 2007.
33. K. Dowlatshahi, M. Fan, H. C. Snider, and F. A. Habib, Lymph node micrometastases from breast carcinoma: Reviewing the dilemma, *Cancer* 80(7), 1188–1197, 1997.
34. H. Kobayashi, S. Kawamoto, M. Bernardo, M. W. Brechbiel, M. V. Knopp, and P. L. Choyke, Delivery of gadolinium labeled nanoparticles to the sentinel lymph node: Comparison of the sentinel node visualization and estimations of intra-nodal gadolinium concentration by the magnetic resonance imaging, *J Control Release* 111(3), 343–351, 2006.
35. R. Jain, P. Dandekar, and V. Patravale, Diagnostic nanocarriers for sentinel lymph node imaging, *J Control Release* 138(2), 90–102, 2009.
36. L. A. Bentolila, Y. Ebenstein, and S. Weiss, Quantum dots for in vivo small-animal imaging, *J Nucl Med* 50(4), 493–496, 2009.
37. K.-T. Yong, I. Roy, H. Ding, E. J. Bergey, and P. N. Prasad, Biocompatible near-infrared quantum dots as ultrasensitive probes for long-term in vivo imaging applications, *Small* 5(17), 1997–2004, 2009.
38. B. Ballou, L. A. Ernst, S. Andreko et al., Sentinel lymph node imaging using quantum dots in mouse tumor models, *Bioconjug Chem* 18(2), 389–396, 2007.
39. B. Bohunicky and S. A. Mousa, Biosensors: The new wave in cancer diagnosis, *Nanotechnol Sci Appl* 4, 1–10, 2011, doi: 10.2147/NSA.S13465.
40. G. Zheng and C. M. Lieber, Nanowire biosensors for label-free, real-time, ultrasensitive protein detection, *Methods Mol Biol* 790, 223–237, 2011, doi: 10.1007/978-1-61779-319-6_18.
41. F. Patolsky, G. Zheng, C. M. Lieber, *Analytical Chemistry*, ACS Publications, Washington, DC, pp. 4261–4269, 2006.
42. S. M. Sze and K. K. Ng, *Physics of Semiconductor Devices*, Wiley, New York, 1981.
43. D.-E. van der Merwe, K. Oikonomopoulou, J. Marshall, and E. P. Diamandis, Mass spectrometry: Uncovering the cancer proteome for diagnostics, PudMed, *Adv Cancer Res* 96, 23–50, 02/2007, doi: 10.1016/S0065-230X(06)96002-3.

44. M. A. Swartz and A. W. Lund, Lymphatic and interstitial flow in the tumour microenvironment: Linking mechanobiology with immunity, *Nat Rev Cancer* 12, 210–219, 2012.

45. F. Michor, J. Liphardt, M. Ferrari, and J. Widom, What does physics have to do with cancer, *Nat Rev Cancer* 11, 657–670, 2011.

46. M. N. Rhyner, A. M. Smith, X. Gao, H. Mao, L. Yang, and S. Nie, Quantum dots and multifunctional nanoparticles: New contrast agents for tumor imaging, *Nanomedicine* 1(2), 209–217, 2006.

47. O. Rabin, J. Manuel Perez, J. Grimm, G. Wojtkiewicz, and R. Weissleder, An X-ray computed tomography imaging agent based on long-circulating bismuth sulphide nanoparticles, *Nat Mater* 5(2), 118–122, 2006.

48. R. Weissleder, D. D. Stark, B. L. Engelstad et al., Super paramagnetic iron oxide: Pharmacokinetics and toxicity, *AJR Am J Roentgenol* 152, 167–173, 1989.

49. U. Ayanthi Gunasekera, Q. A. Pankhurst, and M. Douek, Imaging applications of nanotechnology in cancer, *Target Oncol* 4(3), 169–181, 2009, doi: 10.1007/s11523-009-0118-9.

50. W. J. Mulder, R. Koole, R. J. Brandwijk et al., Quantum dots with a paramagnetic coating as a bimodal molecular imaging probe, *Nano Lett* 6(1), 1–6, 2006.

51. M. Dahan, S. Lévi, C. Luccardini, P. Rostaing, B. Riveau, and A. Triller, Diffusion dynamics of glycine receptors revealed by single-quantum dot tracking, *Science* 302, 442–445, 2003.

52. D. S. Lidke, P. Nagy, R. Heintzmann, D. J. Arndt-Jovin, J. N. Post, H. E. Grecco, E. A. Jares-Erijman, and T. M. Jovin, Quantum dot ligands provide new insights into erbB/HER receptor mediated signal transduction: An overview of liquid chromatography coupled with tandem mass spectroscopy (LC-MS/MS), *Nat Biotechnol* 22, 198–203, 2004.

53. X. Gao, Y. Cui, R. M. Levenson, L. W. K. Chung, and S. Nie, In vivo cancer targeting and imaging with semiconductor quantum dots, *Nat Biotechnol* 22(8), 969–976, 2004, doi: 10.1038/nbt994.

54. J. D. Watrous, T. Alexandrov, and P. C. Dorrestein, The evolving field of imaging mass spectrometry and its impact on future biological research, *J Mass Spectrom* 46(2), 209–222, 2011, doi: 10.1002/jms.1876.

55. K. Schwamborn and R. M. Caprioli, Molecular imaging by mass spectrometry—Looking beyond classical histology, *Nat Rev Cancer* 10, 639–646, 2010.

56. J. J. Pitt, Principles and applications of liquid chromatography–mass spectrometry in clinical biochemistry, *Clin Biochem Rev* 30(1), 19–34, 2009.

57. S. Forcisi, F. Moritz, B. Kanawati, D. Tziotis, R. Lehman, and P. Schmitt-Kopplin, Liquid chromatography–mass spectrometry in metabolomics research: Mass analyzers in ultra-high pressure liquid chromatography coupling, *J Chromatogr A* 1292, 51–65, 2013.

58. M. Mann, R. C. Hendrickson, and A. Pandey, Analysis of proteins and proteomes by mass spectrometry, *Rev Biochem* 70, 437–473, 2001.

59. S. B. Muthu Vadivel, R. Suresh Kumar, A. Tamil Selvan, and R. Suthakaran, An overview of liquid chromatography coupled with tandem mass spectroscopy (LC-MS/MS), Reference Id: Pharmatutor-Art-1581, http://www.pharmatutor.org/articles/overview-liquid-chromatography coupled-tandem-mass-spectroscopy-lc-ms.

60. J. L. Spratlin, N. J. Serkova, and S. G. Eckhardt, Clinical applications of metabolomics in oncology: A review, *Clin Cancer Res* 15(2), 431–440, 2009.

61. P. W. Lee, S. H. Hsu, J. J. Wang et al., The characteristics, bio distribution, magnetic resonance imaging and biodegradability of super paramagnetic core-shell nanoparticles, *Biomaterials* 31, 1316–1324, 2010.
62. Y. Lee and E. D. Baron. Photodynamic therapy: Current evidence and applications in dermatology. *Semin Cutan Med Surg* 30, 199–209, 2011.
63. N. Yavari, S. Andersson-Engels, U. Segersten, and P. U. Malmstrom, An overview on preclinical and clinical experiences with photodynamic therapy for bladder cancer, *Can J Urol* 18, 5778–5786, 2011.
64. C. M. Moore, M. Emberton, and S. G. Bown, Photodynamic therapy for prostate cancer—An emerging approach for organ-confined disease, *Lasers Surg Med* 43, 768–775, 2011.
65. R. Allison, K. Moghissi, G. Downie, and K. Dixon, Photodynamic therapy (PDT) for lung cancer, PubMed, *Photodiagnosis Photodyn Ther*, 8(3), 231–239, September 2011, doi: 10.1016/j.pdpdt.2011.03.342. Epub July 31, 2011.
66. K. K. Wang and J. Y. Kim, Photodynamic therapy in Barrett's esophagus, *Gastrointest Endosc Clin N Am* 13, 483–489, vii, 2003.
67. B. G. Fan and A. Andren-Sandberg, Photodynamic therapy for pancreatic cancer, *Pancreas* 34, 385–389, 2007.
68. J. B. Wang and L. X. Liu, Use of photodynamic therapy in malignant lesions of stomach, bile duct, pancreas, colon and rectum, *Hepatogastroenterology* 54, 718–724, 2007.
69. H. Quon, C. E. Grossman, J. C. Finlay et al., Photodynamic therapy in the management of pre-malignant head and neck mucosal dysplasia and microinvasive carcinoma, *Photodiagnosis Photodyn Ther* 8, 75–85, 2011.
70. P. A. Netti, S. Roberge, Y. Boucher, L. T. Baxter, and R. K. Jain, Effect of transvascular fluid exchange on pressure-flow relationship in tumors: A proposed mechanism for tumor blood flow heterogeneity, *Microvasc Res* 52, 27–46, 1996.
71. R. Tong, H. D. Hemmati, R. Langer, and D. S. Kohane, Photo switchable nanoparticles for triggered tissue penetration and drug delivery, *J Am Chem Soc* 134, 8848–8855, 2012.
72. X. L. Wu, J. H. Kim, H. Koo et al., Tumor-targeting peptide conjugated pH-responsive micelles as a potential drug carrier for cancer therapy, *Bioconjug Chem* 21, 208–213, 2010.
73. Y. L. Colson and M. W. Grinstaff, Biologically responsive polymeric nanoparticles for drug delivery, *Adv Mater* 24, 3878–3886, 2012.
74. C. Wong, T. Stylianopoulos, J. Cui et al., Multistage nanoparticle delivery system for deep penetration into tumor tissue, *Proc Natl Acad Sci USA* 108, 2426–2431, 2011.
75. C. D. Landon, J.-Y. Park, D. Needham, and M. W. Dewhirst, Nanoscale drug delivery and hyperthermia: The materials design and preclinical and clinical testing of low temperature-sensitive liposomes used in combination with mild hyperthermia in the treatment of local cancer, *Open Nanomed J* 3, 38–64, 2011.
76. N. Ochekpe, P. Olorunfemi, and N. Ngwuluka, Nanotechnology and drug delivery Part 1: Background and applications, *Trop J Pharm Res* 8, 265–274, 2009.
77. A. A. Manzoor, L. H. Lindner, C. D. Landon et al., Overcoming limitations in nanoparticle drug delivery: Triggered, intravascular release to improve drug penetration into tumors, *Cancer Res* 72(21), 5566–5575, 2012.

78. R. D. Hofheinz, S. U. Gnad-Vogt, U. Beyer, and A. Hochhaus, Liposomal encapsulated anti-cancer drugs, *Anti-Cancer Drugs* 16(7), 691–707, 2005.

79. J. P. Rao and K. E. Geckeler, Polymer nanoparticles: Preparation techniques and size-control parameters, *Prog Poly Sci* 36(7), 887–913, 2011.

80. C.-M. Hu, S. Aryal, and L. Zhang. Nanoparticles assisted combination therapies for effective cancer treatment, *Ther Deliv* 1, 323–334, 2010.

81. C. Oerlemans, W. Bult, M. Bos, G. Storm, J. F. W. Nijsen, and W. E. Hennink, Polymeric micelles in anticancer therapy: Targeting, imaging and triggered release, *Pharm Res* 27, 2569–2589, 2010.

82. G. C. Giakos, Novel biomolecular nanophotonic devices, sensors, photonic nanocrystals and biochips: Advances in medical diagnostics, environmental and defense, *IEEE International Workshop on Imaging Systems and Techniques—IST*, Krakow, Poland, 1–6, 2007, doi: 10.1109/IST.2007.379578.

83. J. S. Baba, J. R. Chung, A. H. De Laughter, B. D. Cameron, and G. L. Cote, Development and calibration of an automated Mueller matrix polarization imaging system, *J Biomed Opt* 7(3), 341–349, 2002.

84. G. C. Giakos, Multifusion multispectral lightwave polarimetric detection principles and systems, *IEEE Trans Instrum Meas* 55(6), 1904–1912, 2006.

85. G. C. Giakos, V. Adya, K. Valluru, P. Farajipour, S. Marotta, J. T. Paxitzis, M. Becker, P. Bathini, K. Ambadipudi, and S. M. Mandadi, Molecular imaging of tissue with high-contrast multispectral photosensitizing polarimetric imaging techniques, *2009 IEEE International Workshop on Imaging Systems and Techniques (IST '09)*, Shenzhen, China, pp. 63–66, May 11–12, 2009.

86. G. C. Giakos, S. A. Paturi, P. Bathini, S. Sukumar, K. Ambadipudi, K. Valluru, D. Wagenar, V. Adya, and M. Reddy, New pathways towards the enhancement of the image quality, *IEEE IMTC Technical Conference*, Warsaw, Poland, pp. 1–5, May 1–3, 2007, doi: 10.1109/IMTC.2007.379461.

87. G. Giakos, K. Valluru, V. Adya et al., Stokes parameters imaging of multi-index of refraction biological phantoms utilizing optically active molecular contrast agents, *Meas Sci Technol (Special Issue in Imaging Systems and Techniques)* 20, 12–16, 104003, 2009.

88. G. Livanos, M. Zervakis, G. C. Giakos, K. Valluru, S. Paturi, and S. Marotta, Modeling the characteristics of material distributions in polarimetric images, *IET Image Process* 5(5), 429–439, 2011.

89. M. Felton, K. P. Gurton, D. Ligon, and A. Raglin, Discrimination of objects within polarimetric imagery using principal component and cluster analysis, Army Research Laboratory, ARL-TR-4216, 2007.

90. G. C. Giakos, Multifusion, multispectral, optical polarimetric imaging sensing principles, *IEEE Trans Instrum Meas* 55(5), 1628–1633, 2006.

91. R. A. Lewis, K. D. Rogers, C. J. Hall, E. Towns-Andrews, S. Slawson, A. Evans, S. E. Pinder, I. O. Ellis, C. R. M. Boggis, A. P. Hufton, and D. R. Dance, Breast cancer diagnosis using scattered X-rays, *J Synchrotron Rad* 7(Pt. 5), 348–352, 2000.

92. M. Born and E. Wolf, *Principles of Optics*, 7th edn., Cambridge University Press, Cambridge, U.K., 1999.

93. V. Tuchin, *Handbook of Optical Biomedical Diagnostics*, SPIE, Bellingham, WA, 2002.

94. D.-C. Cheng, X. Jiang, and A. Schmidt-Truckass, Image segmentation using histogram fitting and spatial information, P. Perner and O. Salvetti (eds.), *MDA* 2006/2007, LNAI 4826, Springer-Verlag, Berlin, Germany, 2007, pp. 47–57, 2007.

95. X. Zhuang, T. Wang, and P. Zhang, A highly robust estimator through partially likelihood function modeling and its application in computer vision, *IEEE Trans Pattern Anal Mach Intell* 14(1), 19–35, 1992.

96. T. K. Moon, The expectation-maximization algorithm, *IEEE Signal Process Mag*, 13(6), 47–60, 1996.

97. J. Zhang and T. Tan, Brief review of invariant texture analysis methods, *Pattern Recog* 35(3), 735–747, 2002.

98. A. Papoulis and S. U. Pillai, *Probability, Random Variables and Stochastic Processes*, McGraw-Hill Europe, New York, January 1, 2002.

99. Y.-M. F. Goh, H. L. Kong, and C.-H. Wang, Simulation of the delivery of doxorubicin to hepatoma, *Pharm Res* 18, 761–770, 2001.

100. L. T. Baxter and R. K. Jain, Transport of fluid and macromolecules in tumors. I. Role of interstitial pressure and convection. *Microvasc Res* 37, 77–104, 1989.

101. L. T. Baxter and R. K. Jain, Transport of fluid and macromolecules in tumors. II. Role of heterogeneous perfusion and lymphatics, *Microvasc Res* 40, 246–263, 1990.

102. L. T. Baxter and R. K. Jain, Transport of fluid and macromolecules in tumors. III. Role of binding and metabolism, *Microvasc Res* 41, 5–23, 1991.

103. C. S. Teo, W. Hor Keong Tan, T. Lee, and C.-H. Wang, Transient interstitial fluid flow in brain tumors: Effect on drug delivery, *Chem Eng Sci* 60, 4803–4821, 2005.

104. M. Kounelakis, M. Zervakis, G. Giakos, G. Postma, G. L. Buydens, and L. X. Kotsiakis, On the relevance of glycolysis process on brain gliomas, *IEEE J Biomed Health Inform* 17(1), 128–135, 2013.

105. C. Liu, J. Krishnan, and X. Y. Xu, Towards an integrated systems-based modelling framework for drug transport and its effect on tumour cells, *J Biol Eng* 8, 3, 1–19, 2014.

106. S. J. Glover and K. R. Allen, Measurement of benzodiazepines in urine by liquid chromatography-tandem mass spectrometry: Confirmation of samples screened by immunoassay, *Ann Clin Biochem* 47(2), 111–117, 2010.

107. M. Johannsen, A. Jordan, R. Scholz et al., Evaluation of magnetic fluid hyperthermia in a standard rat model of prostate cancer, *J Endourol* 18, 495–500, 2004.

108. A. Jordan, R. Scholz, K. Maier-Hauff et al., The effect of thermotherapy using magnetic nanoparticles on rat malignant glioma, *J Neurooncol* 78, 7–14, 2006.

109. A. Jordan, R. Scholz, P. Wust et al., Effects of magnetic fluid hyperthermia (MFH) on C3H mammary carcinoma in vivo, *Int J Hyperthermia* 13, 587–605, 1997.

110. F. Alexis, E. M. Pridgen, R. Langer, and O. C. Farokhzad, Nanoparticle technologies for cancer therapy, *Handb Exp Pharmacol* (197), 55–86, 2010.

111. K. K. Jain, *The Handbook of Nanomedicine*, Humana Press, NJ, 2008, ISBN: 1603273182.

112. F. Ge and Q.-Y. He, Proteomics in cancer biomarker discovery, X. Zhang (ed.), *Omics Technologies in Cancer Biomarker Discovery*, pp. 39–48, Chapter 3, Landes Bioscience, Austin, TX, 2011.

113. S. L. Liang and D. W. Chan, Enzymes and related proteins as cancer biomarkers: A proteomic approach, *Clin Chim Acta* 381, 93–97, 2007.

114. E. M. Posadas, F. Simpkins, L. A. Liotta et al., Proteomic analysis for the early detection and rational treatment of cancer—Realistic hope, *Ann Oncol* 16(1), 16–22, 2005.

115. M. Latterich, M. Abramovitz, and B. Leyland-Jones, Proteomics: New technologies and clinical applications, *Eur J Cancer* 44(18), 2737–2741, 2008.
116. W. C. Cho and C. H. Cheng, Oncoproteomics: Current trends and future perspectives, *Expert Rev Proteomics* 4(3), 401–410, 2007.
117. A. Rouxa, D. Lisonb, C. Junota, and J.-F. Heiliera, Applications of liquid chromatography coupled to mass spectrometry-based metabolomics in clinical chemistry and toxicology: A review, *Clin Biochem* 44(1), 119–135, 2011.
118. I. Ali, Z. A. Al-Othman, K. Saleem, A. Hussain, and I. Hussain, Role of chromatography for monitoring of breast cancer biomarkers, *Recent Patents Biomarkers* 1, 89–97, 2011.
119. S. Forcisi, F. Moritz, B. Kanawatia, D. Tziotis, R. Lehmann, and P. Schmitt-Kopplin, Liquid chromatography–mass spectrometry in metabolomics research: Mass analyzers in ultrahigh pressure liquid chromatography coupling, *J Chromatogr A* 1292, 51–65, 2013.

6

Nanoscale Flow Application in Medicine

Viroj Wiwanitkit

CONTENTS

6.1 Introduction 172
6.2 Overview of Flow and Its Application in Medicine 175
6.3 Computational Flow Analysis in Medicine 182
    6.3.1 Usefulness of Computational Flow Analysis in Medicine 182
    6.3.2 Summary of Important Reports on Applications
        of Computational Flow Analysis in Medicine 184
    6.3.3 Important Computational Flow Analysis
        Tools for Biomedical Work 185
6.4 Microflow and Nanoflow in Medicine 185
    6.4.1 Usefulness of Microflow and Nanoflow in Medicine 185
    6.4.2 Microfluidics and Nanofluidics in Medicine 186
    6.4.3 Summary of Important Reports on Microfluidics
        and Nanofluidics 187
        6.4.3.1 Reports on Microfluidics 187
        6.4.3.2 Reports on Nanofluidics 188
    6.4.4 Important Computational Tools for Microflow and
        Nanoflow 188
6.5 Common Applications of Computational Nanoflow in
    General Medical Practice 189
    6.5.1 Examples of Medical Research Based on
        Computational Nanoflow Application 191
        6.5.1.1 Example 1 191
        6.5.1.2 Example 2 192
        6.5.1.3 Example 3 193
        6.5.1.4 Example 4 194
6.6 Conclusion 195
References 196

6.1 Introduction

Flow is an important term in science. When one mentions flow, it usually means something related to dynamicity. Focusing on the material things in our world, there are three states: solid, fluid, and gas. A solid is a constant with a formed element and shape; it is visible. A fluid is labile and does not have any shape of its own. It can be given a shape by a solid container. A fluid is visible. A gas is another state; it is not visible. Similar to a fluid, a gas is labile and does not have any shape of its own. It can also be given a shape by a solid container. Focusing on the term *fluid*, it involves two phases: liquid and gas.

Flow also gives the sense of *moving*. Talking about *flow*, one can relate to movement and motility. Indeed, flow is dynamic and can relate to the movement of molecules and further to *dynamic energy*. It is not possible to talk about flow without referring to movement. As already mentioned, flow is directly related to fluid and gas. It can be simply said that flow is the basic property of any fluid and gas. This physical property can be mentioned and classified as a specific property of a *fluid*. So any liquid or gas is also known as a fluid. And the science that deals with fluids is called fluid science.

At present, *flow* is not a new concept. It is actually a classical concept in science that is still important today. The scientific society turns its current focus to more in-depth studies of flow properties. The physical principle of flow poses challenges to science. This leads to the consideration of the application of flow in the scientific world. In the past, the limitations of human ability restricted the study of flow. However the development of new theories, especially in computer science, presents the scientific society with a new facet for the application of flow. With the use of this new scientific tool, flow can be easily manipulated.

This is a jump to a new paradigm that deals with *flow* in science. Many new concepts have been introduced over the years. Nevertheless, scientists continue their research and find new aspects in the field of *flow*. The science of flow falls within the scope of physics. The history of flow physics is very long and interesting. The two states of liquid and gas are mainly dealt with in flow physics. Flow physics is a branch of physics that is very useful. It can be integrated with computational technology. Flow physics is the basic requirement and knowledge for mechanical engineering. The advanced knowledge of flow physics contributes to many new theories and models. At present, many new computational tools and laboratory techniques are introduced in flow physics. The advent of many new concepts in flow physics allow scientists to successfully conduct fundamental, experimental, and computational research. This can help enhance the knowledge and understanding of the physics underlying boundary-layer transition, turbulence, and vortical and separated flows. This can also help lead to accurate engineering design analysis and control of complex flows. At present, flow is studied worldwide. Many specific scientific societies are set corresponding to *flow* science. There are many research groups in this field such as the group at Stanford University.

Currently, scientists successfully manipulate extremely small scales at the nanolevel, and this is the root of a new science, namely, nanoscience. At this level, the extremely small scale is called nanoscale. At present, new science focuses and acts on this nanolevel; hence, the *fluid* can be a topic directly dealt with in nanoscience. This means that scientists currently deal with nanoscale fluid. This new approach is a real challenge in the study of fluid. At present, nanoscience has wide applications; hence, there is no doubt that it can be applied to fluid science. In fact, nanoscience can be useful in understanding *fluid* and *flow*. The application of nanoscience for *fluid* and *flow* can be seen in many aspects. With the use of nanoscience, many new facts of *fluid* and *flow* can be discovered. Many secrets can come to light and become part of new knowledge. As already mentioned, nanoscience guides the study of nanoscale phenomena, which include *fluid* and *flow*. Fluid theory is the core theory that describes the physical property of a fluid molecule as a small dynamic object. The fluid molecule can be studied and explained in nanoscience, and its dynamics can be explained by the principles of any object in the nanolevel scale. However, there are some interesting properties of fluid in terms of nanoscale objects that should be mentioned.

Basically, when an object decreases to a very small size, there are new characteristics that emerge. When the size of an object changes, it refashions itself. Its new different properties can be seen, and this is the basic rule in modern nanoscience. The differences can be seen in both physical and chemical characteristics. The new properties of nanoobjects make them not totally the same as their corresponding large-scale form. First, when an object is changed into a very small size, significant changes in its surface area can be expected and this leads to many additional changes (such as electricity). There is no doubt that the advanced computational and electronic tool can be applied for dealing with fluids. This has become the new concept and facet in mechanical engineering.

In fact, nanoscience is not strictly assigned to biological or physical aspects. Hence, nanoscience can act as an interdisciplinary bridge between physical and biological sciences. It can be classified as a hybrid between physical and biological sciences. The work in physical, biological, or hybrid fields can be seen within the scope of nanoscience. In fact, nanoscience is studied worldwide and can manipulate situations with complex small molecules. Molecules of fluids and gases are usually small and in visible. Hence, the application of nanoscience helps deal with those extraordinarily small molecules. It helps resolve questions on the flow of extremely small objects, including gas and fluid molecules. Questions that were previously difficult to solve can be simply answered based on nanoscience. Manipulation based on nanoscience becomes an effective tool for scientists who seek to answer complex questions. Hence, nanoscience can play an important role as a new alternative approach for solving the problem of *flow*.

As already mentioned, flow science deals with several aspects of fluid. The topics to deal with include basic physics, acoustics, transition, heat transfer, thermodynamics, chemical reactions, biofluids, complex fluids, fluidics, and plasmas. Several phenomena are involved, such as aerodynamics, electronics cooling, environmental engineering, materials processing, planetary entry, propulsion, and power systems. To complete the work of flow science, there is a requirement for new effective methods and tools for generation, access, display, interpretation, and postprocessing of available databases on fluids. Basic knowledge in many sciences including acoustics, aerodynamics, computational fluid mechanics, computational mathematics, fluid mechanics, combustion, thermodynamics, and propulsion is needed.

To facilitate nanoscience work, whether in vivo, in vitro, or in silico, (computational) techniques can be selected for and application [1]. Classically, the two common techniques in medicine are in vivo and in vitro. However, these two classical approaches require a lot of material to perform the actual tests in human or laboratory conditions. This involves risk and high cost. The latest concept is the use of computational modeling, which is widely known as the in silico alternative. The in silico alternative is a widely applied method in computational medicine at present. It is the focused interest of the scientific society. In nanoscience too, the in silico alternative can be used in addition to the two classical approaches of in vivo and in vitro techniques. This helps overcome the basic limitations of in vivo and in vitro approaches, which include high cost and time consumption. Hence, it is not surprising to find the use of the in silico approach in nanomedicine. This new approach can help clarify and manipulate many research questions and developments in nanoscience.

Computational technology can be applied worldwide at present. It can be used as in silico approach in science including nanoscience. The so-called computational medicine is an interesting field of biomedical sciences. The effectiveness and reliability of computational technology have been confirmed. The latest in silico technique, nanoinformatics, which deals with the nanolevel, can additionally support the classical in vivo and in vitro experiments. Applying this technique, a significant reduction in terms of time and cost can be expected. As already mentioned, computational nanomedicine can give solutions to difficult questions in nanomedicine. This is feasible within a short time at a low unit cost. With the use of simulation, confounding factors or interferences can be resolved. The in silico approach is free of interference, which is a positive aspect compared to real studies.

Focusing on the in silico approach, the blooming of the *omics* sciences should be mentioned [2]. It is widely used around the world at present. Several new *omics* sciences have emerged as interesting scientific techniques that can be widely applied. Similar to any field of scientific study, nanomedicine can apply *omics* in its usage. With the help of available omics techniques, solutions to many complex problems can be simply attained. In silico techniques can be useful in giving fast answers to several complex nanoscale problems. Clarification of a scenario or prediction of an imaginary phenomenon, which are common complex problems in classical medicine, can be done [3–5].

In medicine, the study of fluids is widely conducted. Body fluid is the king of fluids; hence, the application of flow science is possible. Indeed, the topic of the flow of fluid is not a new concept but a very old topic in medicine. The properties of body fluid have been studied in medicine for a very long time. There is a well-known medical science that deals with fluid known as medical technology. Medical technology is within the scope of medical science that is widely studied and is generally introduced in a medical society and becomes the basic required knowledge for all physicians and medical personnel. For many years, this science has been included in the biomedical curriculum.

As noted, the study of the properties of body fluid is very complex and requires high technology. Many new advanced tools, including medical microscopes, are available for medical technologists to determine the properties of body fluid. Sometimes, one has to deal with very small molecules within biofluids by observing or measuring them, for example. Novel nanomedicine techniques can no doubt be applied for such studies in medical technology. This can be done by the specific branch of nanomedicine that makes use of computational application called nanoinformatics.

The imaginary in silico computational approach in nanomedicine by a specific technique called nanoinformatics can be used elsewhere as well. Several data can be well manipulated for application purposes in nanoinformatics. This specific subbranch of nanomedicine is a modern advent that can be very useful in medicine. Another important application of nanomedicine is the applied fluids technology. This can be in the form of either microfluidics or nanofluidics. In medical physics, microfluidics and nanofluidics are new approaches with complex techniques. Novel techniques based on nanofluids are also introduced for use in medicine, the details of which are given in the following sections.

6.2 Overview of Flow and Its Application in Medicine

A fluid is a visible matter in medicine. One cannot overlook the main property of medicine, its flow, when talking about medical fluid. *Fluid* and *flow* have been mentioned in biomedical sciences for a long time. Several fluids have been introduced for clinical correlation in medicine. For centuries, fluid has been known, examined, and applied for clinical studies in biomedicine. Fluid interaction is an important topic that is widely studied in medicine.

To deal with fluid interaction, specific knowledge of medical technology is necessary. But first, *fluid dynamics* should be understood.

Fluid dynamics is an important subject in general physics. As an important property of fluid is flow, fluid dynamics mainly deals with flow. In dynamics, motion is the first operative function. Fluid dynamics can be either aerodynamics for gas or hydrodynamics for liquid. The principles of fluid dynamics mainly cover the measurement of flow and its related phenomena. Various properties of fluids including velocity, pressure, density, and temperature are usually dealt with. The study can be based on the mathematical model based on mathematical functions of space and time.

To deal with fluid dynamics, several mathematical equations have to be considered.

1. *Conservation laws*: The basic rules of fluid dynamics are conservation laws. These laws include conservation of mass, conservation of linear momentum, and conservation of energy. These laws confirm the property of a fluid as a substance. There are four main points to be noted. First, a fluid has a continuum of mass. This means mass continuity. Second, a fluid has conservation of momentum. Third, a fluid has conservation of energy. The manipulation of fluid molecules at an extremely small scale is possible.

2. *Compressible*: Whether a fluid is compressible or not is an interesting question. In fact, all fluids are compressible but only to a limit. When it is compressed, the changes in pressure and temperature can be seen. In addition, the consequent changes in density can be observed. During any flow, compression can be expected. This is also related to pressure and temperature.

3. *Viscosity*: Viscosity is an important property of any fluid. The Reynolds number, which is the ratio between inertial and viscous forces, can be assigned to any fluid. This knowledge is the basic requirement for the study of any fluid's viscosity.

   Basically, when one talks about viscosity, two forces should be noted: inertial force and viscous force. In the case of a low Reynolds number, viscous force plays a main role; on the other hand, for a high Reynolds number, inertial force plays a main role.

4. *Steady*: When the time derivatives of flow do not exist, it is said to be a steady flow. It can be simply said that a steady-state flow is a condition where the fluid properties at a point in the system do not change at any time.

5. *Laminar and turbulent*: The flow can be changeable. When recirculation occurs, it means there is turbulence. On the other hand, when there is no recirculation, it is called a laminar flow.

6. *Magnetohydrodynamics*: This is the specific description of a fluid in a multifaceted approach. The corresponding approach in the field of electromagneticity has been established. Therefore, the complex approach has to be used in magnetohydrodynamics.

In science, the application of fluid dynamics can be seen in several ways. This also includes the application in medicine. To help the reader better understand, applications to the aforementioned mathematical equations are shown as follows.

1. *Conservation laws*: The following publications are good examples of conservation laws of fluid in medicine:
   a. Balankin and Elizarraraz reported on a map of fluid flow in fractal porous medium into fractal continuum flow [6].
   b. Sbragaglia and Sugiyama studied a volumetric formulation for a class of kinetic models with energy conservation [7].
   c. Soltani and Chen reported on the numerical modeling of fluid flow in solid tumors [8].
2. *Compressible*: The following publications are good examples of the compressible property of fluid in medicine:
   a. Guédra et al. reported on ultrasonic propagation in suspensions of encapsulated compressible nanoparticles [9].
   b. Cuellar et al. reported on a novel approach for measuring the intrinsic nanoscale thickness of polymer brushes by means of atomic force microscopy [10]. This work can be a good example of an application of a compressible fluid model [10].
   c. Tatsumi and Yamamoto reported on velocity relaxation of a particle in a confined compressible fluid [11].
   d. Brenner proposed a critical test of the Navier–Stokes–Fourier paradigm for compressible fluid continua [12].
   e. Gross and Varnik reported on critical dynamics of an isothermal compressible nonideal fluid [13].
   f. Godin reported on the incompressible wave motion of compressible fluids [14].
   g. Felderhof reported on dissipation in peristaltic pumping of a compressible viscous fluid through a planar duct or a circular tube [15].
   h. Ebin reported on the motion of a slightly compressible fluid [16].
   i. Franken et al. reported on the oscillating flow of a viscous compressible fluid through a rigid tube [17].

j. Herlofson reported on magneto-hydrodynamic waves in a compressible fluid conductor [18].

k. Staroselsky et al. reported on the long-term, large-scale properties of a randomly stirred compressible fluid [19].

3. *Viscosity*: The following publications are good examples of the viscosity of fluids in medicine:

a. Kim et al. reported on the viscosity of magnetorheological fluids using iron–silicon nanoparticles [20].

b. Xu discussed how to decrease the viscosity of a suspension with a second fluid and nanoparticles [21].

c. Eddi et al. reported on the influence of droplet geometry on the coalescence of low-viscosity drops [22].

d. Hoang and Galliero reported on the local shear viscosity of strongly inhomogeneous dense fluids: from the hard sphere to the Lennard–Jones fluids [23].

e. Ishikawa et al. reported a study on EGG affected by the viscosity and shear rate dependence of fluids and semisolid diets [24].

f. Prakash et al. reported on the interaction of bubbles in an inviscid and low-viscosity shear flow [25].

g. Jestrović et al. reported on the effects of increased fluid viscosity on swallowing sounds in healthy adults [26].

h. Holsworth et al. reported on the effect of hydration on whole blood viscosity in firefighters [27].

i. Gachelin et al. reported on the non-Newtonian viscosity of *Escherichia coli* suspensions [28].

j. Llovell et al. reported on the free-volume theory coupled with soft-SAFT for viscosity calculations [29]. In this work, Llovell et al. also performed a comparison with molecular simulation and experimental data [29].

k. Cheng et al. reported on including fluid shear viscosity in a structural acoustic finite element model using a scalar fluid representation [30].

4. *Steady*: The following publications are good examples of the steady state of fluids in medicine:

a. Zhang et al. reported a hemodynamic analysis of renal artery stenosis using computational fluid dynamics technology based on unenhanced steady-state free precession magnetic resonance angiography [31].

b. Throckmorton et al. reported a steady and transient flow analysis of a magnetically levitated pediatric VAD [32].

   c. Ghatage and Chatterji reported a modeling steady-state dynamics of macromolecules in an exponentially stretching flow using multiscale molecular dynamics–multiparticle collision simulations [33].

   d. Martinez et al. reported on monitoring steady flow effects on cell distribution in engineered valve tissues by magnetic resonance imaging [34].

   e. Shehzad et al. reported on the hydromagnetic steady flow of a Maxwell fluid over a bidirectional stretching surface with prescribed surface temperature and prescribed surface heat flux [35].

   f. Larsen and Riisgård reported the validation of the flow-through chamber and steady-state methods for clearance rate measurements in bivalves [36].

   g. Bedkihal et al. reported on a steady flow through a constricted cylinder by multiparticle collision dynamics [37].

5. *Laminar and turbulent*: The following publications are good examples of laminar and turbulent properties of a fluid in medicine:

   a. Potvin reported on hydrodynamics of the turbulent point-spread function [38].

   b. Pravin and Reidenbach reported that simultaneous sampling of flow and odorants by crustaceans can aid searches within a turbulent plume [39].

   c. Bourgogne et al. reported on rugged and accurate quantitation of colchicine in human plasma to support colchicine poisoning monitoring by using turbulent-flow LC-MS/MS analysis [40].

   d. Kwon et al. reported the temperature dependence of convective heat transfer with Al_2O_3 nanofluids in the turbulent flow region [41].

   e. Skartlien et al. reported on droplet size distributions in turbulent emulsions [42].

   f. Yadav and Bhagoria reported on modeling and simulation of turbulent flows through a solar air heater having square-sectioned transverse rib roughness on the absorber plate [43].

   g. Shetty and Frankel reported an assessment of stretched vortex subgrid-scale models for LES of incompressible inhomogeneous turbulent flow [44].

   h. Wu et al. reported on Von Kármán energy decay and heating of protons and electrons in a kinetic turbulent plasma [45].

   i. Kim and Sureshkumar reported on the spatiotemporal evolution of hairpin eddies, Reynolds stress, and polymer torque in polymer drag-reduced turbulent channel flows [46].

   j. Nootz et al. reported the determination of flow orientation of an optically active turbulent field by means of a single beam [47].

   k. Lee et al. reported on evaluating the performance of a turbulent wet scrubber for scrubbing particulate matter [48].

   l. Klinkenberg et al. reported a numerical study of laminar–turbulent transition in a particle-laden channel flow [49].

6. *Magnetohydrodynamics*: The following publications are good examples of magnetohydrodynamics of a fluid in medicine:

   a. Sahore and Fritsch reported on flat flow profiles achieved with microfluidics generated by redox magnetohydrodynamics [50].

   b. Khan et al. reported on Stokes' second problem for magnetohydrodynamics flow in a Burgers' fluid [51].

   c. Arter reported on a potential vorticity formulation of compressible magnetohydrodynamics [52].

   d. Gao et al. reported on 3D imaging of flow patterns in an internally pumped microfluidic device, redox magnetohydrodynamics, and electrochemically generated density gradients [53].

   e. Morales et al. reported the intrinsic rotation of toroidally confined magnetohydrodynamics [54].

   f. Ensafi et al. reported the redox magnetohydrodynamics enhancement of stripping voltammetry of lead(II), cadmium(II), and zinc(II) ions using 1,4-benzoquinone as an alternative pumping species [55].

   g. Krstulovic et al. reported Alfvén waves and ideal 2D Galerkin truncated magnetohydrodynamics [56].

   h. Lyutikov and Hadden reported relativistic magnetohydrodynamics in one dimension [57].

   i. Cheah et al. reported the evaluation of heart tissue viability under redox-magnetohydrodynamics conditions: toward fine-tuning flow in biological microfluidics applications [58].

   j. Onofri et al. reported the effects of anisotropic thermal conductivity in magnetohydrodynamics simulations of a reversed-field pinch [59].

   k. Benzi and Pinton reported the magnetic reversals in a simple model of magnetohydrodynamics [60].

   l. Nigro and Carbone reported the magnetic reversals in a modified shell model for magnetohydrodynamics turbulence [61].

   m. Chatterjee and Amiroudine reported a lattice kinetic simulation of nonisothermal magnetohydrodynamics [62].

n. Mininni and Pouquet reported on finite dissipation and intermittency in magnetohydrodynamics [63].

o. Lessinnes et al. reported on energy transfers in shell models for magnetohydrodynamics turbulence [64].

p. Lee et al. reported on paradigmatic flow for small-scale magnetohydrodynamics: properties of the ideal case and the collision of current sheets [65].

q. Ohkitani and Constantin reported 2D and 3D magnetic reconnection observed in the Eulerian–Lagrangian analysis of magnetohydrodynamics equations [66].

r. Servidio et al. reported the ergodicity of ideal Galerkin 3D magnetohydrodynamics and Hall magnetohydrodynamics models [67].

s. Lu et al. reported on a numerical algorithm for magnetohydrodynamics of ablated materials [68].

t. Simakov and Chacón reported on a quantitative, comprehensive, analytical model for magnetic reconnection in Hall magnetohydrodynamics [69].

u. Chacón et al. reported the steady-state properties of driven magnetic reconnection in 2D electron magnetohydrodynamics [70].

v. Kim reported a consistent theory of turbulent transport in 2D magnetohydrodynamics [71].

w. De Moortel reported on propagating magnetohydrodynamics waves in coronal loops [72].

x. Wheatley et al. reported the stability of an impulsively accelerated density interface in magnetohydrodynamics [73].

y. Graham et al. reported the cancellation exponent and multifractal structure in 2D magnetohydrodynamics: direct numerical simulations and Lagrangian averaged modeling [74].

Based on the examples given, there is no doubt that *flow* can be applied to medicine, especially medical technology. Medical technology is the basic medical science that directly deals with investigation. In fact, this technology is accepted as a simple, accurate, and basic analytical method that can be applied for assessing patients. Several tools based on *flow* are available. In addition, *flow* has its property of dynamics that is concordant with the situation within the human body. Flow exists everywhere in the human body. The best example is the flow within the vessels. Hence, *flow* is the focus of interest in many fields of medicine such as the following:

- *Vascular medicine*: Vascular medicine is the specific medical science that deals with blood vessels. Blood vessels comprise the specific tract for the flow of the most important body fluid, blood. Hence, vascular medicine is the specific medical science that deals with flow.
- *Cardiology*: To assess a cardiovascular disorder, the assessment of flow in cardiac vessels is needed. The flow in blood vessels also relates to the pressure, and this is the clinical correlation to blood pressure.
- *Nephrology*: The blood flow to the kidney is the main concern when one deals with nephrological disorders.
- *Urology*: The urine flow is the important aspect to be assessed in urology. Urological obstruction is a common problem that can be seen worldwide.
- *Neurology*: To assess a neurovascular disorder, the assessment of flow in neurovascular vessels is needed. In addition, the flow in the cerebrospinal fluid tract may also be of concern.
- *Obstetrics*: The assessment of blood flow in the fetus in utero is widely performed in obstetrics. This is a useful tool for monitoring fetal health.
- *Oncology*: The blood flow to the tumor is the main concern in oncology. This usually relates to the growth and metastasis of cancer.
- *Pulmonary medicine*: The flow of air within the airway is the main focus in pulmonary medicine. This is an issue of medicinal flow.
- *Radiology*: There are many new radiological investigations aiming at tracing flow. This can be useful for diagnostic purposes of many disorders.
- *Surgery*: The assessment of flow is the key issue in any surgical procedure, as *flow* usually means blood supply. To cut blood supply, stop bleeding, and reservation of perfusion are the basic requirements in any surgery.

6.3 Computational Flow Analysis in Medicine

6.3.1 Usefulness of Computational Flow Analysis in Medicine

There are several advantages of applying computational technology to medicine, including the determination and assessment of medical flow. Regarding the specific manipulation techniques of flow, a nanoscale approach can be applied when the focused determinant is extremely small. To make successful assessments, a bridging between physical and biological sciences must be done, which leads to the necessity of computer application. Basically, flow is a totally dynamic situation, which concerns both time and site. Hence, this interrelation has a specific site of interaction, which can be called a specific space. Therefore, the concept of the finite element method may be applied. Here, the space can be gridded according to the focus. Additionally, any interaction in flow involves time, as already mentioned. Hence, in a focused interrelationship, both space and time can be completely studied. Theoretically, the time-dependent Maxwell's equations can be used, and this is why the finite difference time domain method can be applied. Space-based central difference estimations and time-based partial derivatives can be selected. With the use of advanced computational technology and the appropriate software and hardware, solving the generated finite difference equations is feasible.

Focusing on flow, it is possible to simulate the interaction of nano-objects in the flow line. The manipulation of nanoscale particles of nano-objects using nanoscience techniques is possible. This can also be done by nanoinformatics, a specific subject on the application of computational techniques in nanoscience. With the use of this nanoscale technique, the flow can be interpreted and computational technology can play a role in this. In medicine, good examples of computational flow analysis can be seen in clinical microscopy, cardiology, neurology, radiology, pulmonary medicine, oncology, and laboratory medicine (Table 6.1). This will be further discussed in the following sections.

TABLE 6.1

Medical Disorders for Which Applied Medical Flow Analysis Can Be Useful in Medicine

Scopes	Examples of Medical Disorders
Cardiology	Myocardial infarction
	Hypertension
	Aneurysm
	Myocardial ischemia
Nephrology	Renal ischemia
	Renal failure
	Renal infarction
	Renal stone
Urology	Ureter stone
	Urethra stone
Neurology	Brain infarction
	Thrombosis
	Hydrocephalus
Pulmonary medicine	Asthma
	Aspiration
	Suffocation
Obstetrics	Feto-fetal transfusion
	Placental insufficiency

6.3.2 Summary of Important Reports on Applications of Computational Flow Analysis in Medicine

As already noted, the application of computational flow analysis can be seen in several areas in medicine. Laboratory medicine is the best example where the application is widely used for clinical microscopy. Clinical microscopy has been introduced in the medical society for a very long time and is one of the core subjects of laboratory medicine. It is the specific subject that deals with the detection of extremely small particles, including those that flow in the stream of body fluid. Since these particles are so small as to be invisible to the naked eye, clinical microscopy techniques help in their assessment. Hence, the application of computational technology in visualization and detection apparatuses can be seen.

As already mentioned, the new microscopic techniques can be used for determination and computational application is possible. The analysis of flowing particles requires an applied microscopic technique that can catch up with their dynamicity. The analysis is directly based on both basic microscopic technology and additional determination techniques (such as staining). The applied computational technology helps manipulation of the derived primary data. To help the reader better understand this application, some interesting reports can be found in the following references:

- Hatami et al. reported computer simulation of MHD blood conveying gold nanoparticles as a third-grade non-Newtonian nanofluid in a hollow porous vessel [75].
- Voronov et al. reported simulation of intrathrombus fluid and solute transport using in vivo clot structures with single platelet resolution [76].
- Dong et al. reported hemodynamics analysis of patient-specific carotid bifurcation: a CFD model of downstream peripheral vascular impedance [77].
- Omodaka et al. reported on local hemodynamics at the rupture point of cerebral aneurysms determined by computational fluid dynamics analysis [78].
- Naito et al. reported on magnetic resonance fluid dynamics for intracranial aneurysms—comparison with computed fluid dynamics [79].
- Qian et al. reported the risk analysis of unruptured aneurysms using computational fluid dynamics technology: preliminary results [80].
- Fukuda et al. reported on computer-simulated fluid dynamics of arterial perfusion in extracorporeal circulation: from reality to virtual simulation [81].
- Wood et al. reported on combined MR imaging and CFD simulation of flow in the human descending aorta [82].

6.3.3 Important Computational Flow Analysis Tools for Biomedical Work

There are several computational flow analysis tools for the management of biomedical data that are listed as follows:

1. *Mytoe*: Mytoe is an automatic analysis of mitochondrial dynamics that includes flow within mitochondria [83].

2. *FAAST*: FAAST is a computational flow-space assisted alignment search tool [84].

3. *flowPhyto*: flowPhyto is a computational tool that helps analyze microscopic algae from continuous flow cytometric data [85].

4. *iFlow*: iFlow is a computational graphical user interface for flow cytometry tools in bioconductors [86].

5. *ITM Probe*: ITM Probe is a computational tool that can analyze information flow in protein networks [87].

6. *COMKAT*: COMKAT is a compartment model kinetic analysis tool [88].

7. *CytoITMprobe*: CytoITMprobe is a computational network information flow plugin for Cytoscape [89].

6.4 Microflow and Nanoflow in Medicine

6.4.1 Usefulness of Microflow and Nanoflow in Medicine

As already mentioned, medical flow is an important subject in medicine. Flow can be seen at all levels. It can manifest as a macrophenomenon as in overt hemorrhage and bleeding. It can also be seen at the micro- and nanolevels. To study and manage the macroflow in medicine is easier than micro- and nanoflows as the naked eye is sufficient for the assessment of the macroflow. However, micro- and nanoflows have many important aspects in medicine. Many disorders occur due to problems in micro- and nanoflows (Table 6.2); hence, it is needed to understand them. The knowledge of micro- and nanoflows in medicine can be useful in both diagnosis and treatment. However, the widely used application is usually in the scope of diagnostic medicine. There are many new tools that manipulate medical micro- and nanoflows that are widely known as microfluidics and nanofluidics. As these tools are usually used for determining extremely small particles within the blood (especially blood cells and platelets), they are routinely used to study *flow cytometry* in medicine. The specific details of these tools will be further discussed in the following section.

TABLE 6.2

Some Important Disorders due to Problems
of Micro- and Nanoflows in Medicine

Group	Examples
Microflow	Capillary leakage
	Petechiae
	Infarction
Nanoflow	Cellular apoptosis
	Mutagenesis

6.4.2 Microfluidics and Nanofluidics in Medicine

Microfluidics and nanofluidics are the tools that are widely used in diagnostic medicine at present. These tools function on the concept of a diagnostic chip. Historically, the technology was developed to help reduce the required workspace for the diagnosis by classical laboratory instruments. Generally, in medicine, the most widely tested samples are blood samples. How to manipulate blood samples within a limited space is the question that medical scientists are usually faced with. As a rule, in vivo monitoring of many important blood parameters is the aim. This is to replace the classical technique with off-line laboratory determination or on-site point-of-care testing. The use of electronics and computational technologies can help consolidate the classical tools, and this is the starting point of laboratory chip development [90].

For any analysis, a small amount of blood is still required. The first success is in manipulating the microlevel of a blood sample. The solution for such an analysis is with microfluidics, which controls the flow of blood samples within a small diagnostic tool. The control of microfluidics in a device is very challenging and is the key issue that determines the success of a microfluidics tool. The flow principle can be explained as a valveless, electroosmotically driven technology [91]. It is used for controlling the stream profile in a laminar flow chamber [91]. Basically, a blood sample is forced to pass through three basic steps within a small bioanalyzer: (1) sample loading, (2) flowing for analysis, and (3) expulsion of an already used sample [91]. The adjustment of the flow of an electroosmotically controlled guiding stream has to be done for positioning the sample stream perpendicular to the flow direction [91]. The capillary force is used for sample loading into the analyzer units [92]. To control the time duration of a sample fluid in each compartment of the device (inlet port, diffusion barrier, reaction chamber, flow-delay neck, and detection chamber), the fluid conduit is used [92]. The conduit has to be well designed with various specifications of channel width, depth, and shape [92]. The fluid path, especially, must be well designed to allow the passage of the sample, and managing naturally stops after filling the detection chamber [92]. Then the biochemical reaction and subsequent washing steps will be conducted at the proper time [92]. The feasibility of creating nanometer-scale depressions in biological substrates using active enzymes delivered with scanning-probe microscopes is the main determinant for the effectiveness of the tool [93]. A well-designed tool must have a good, accurate, and precise control of fluid flow dimension [93]. The same concept is used in nanofluidic biochip applications [93].

Similar to microfluidics, nanofluidics is another important novel technology for developing a diagnostic device. For nanofluidics, the control of a nanovolume of fluid is the focus instead of a microvolume. The two most widely used techniques for nanostructuring and nanochannel fabrication for nanofluidics are the top-down and bottom-up methods [94]. The top-down method is based on patterning on a large scale while reducing the lateral dimensions to the nanoscale [94]. The other approach is the bottom-up method, which arranges molecules in nanostructures [94]. However, the application of nanofluidics is still less than that of microfluidics.

6.4.3 Summary of Important Reports on Microfluidics and Nanofluidics

There are many recent reports on microfluidics and nanofluidics in medicine. Some important reports are listed as follows:

6.4.3.1 Reports on Microfluidics

- Lai et al. presented an integrated microfluidic device for performing an enzyme-linked immunosorbent assay for rat IgG from a hybridoma cell culture [95]. They found that the microchip-based enzyme-linked immunosorbent assay had the same detection range as the conventional method [95].

- Morier et al. studied the microfluidics system and mathematically supported that tilting a 1 cm long covered microchannel was enough to generate flow rates up to 1000 nL/min [96].

- Herrmann et al. described a new microfluidic design with magnetic bead application for the realization of parallel enzyme-linked immunosorbent assay in stop-flow conditions [97].

- Eteshola and Balberg reported that the extension of a hybrid integrated technique to on-chip imaging was possible [98]. They also reported on the quantification of light emission from a biochemical immunoassay in a PDMS chip [98].

- Ding et al. reported on surface acoustic wave microfluidics [99].

- Jin et al. reported a microfluidics platform for a single-shot dose–response analysis of chloride channel-modulating compounds [100].

- Liu et al. reported on microfluidics for the synthesis of peptide-based PET tracers [101].
- Colace et al. discussed the role of microfluidics in coagulation biology [102].
- Al-Hetlani discussed forensic drug analysis by microfluidics tools [103].

6.4.3.2 Reports on Nanofluidics

- Parikesit et al. reported an analysis of electroosmotic flow in a branched-turn nanofluidic device that was specifically produced for the detection and sorting of single molecules [104]. They concluded that the deviation between the measured and simulated data could be explained by the measured Brownian motion of the tracer particles [104].
- Wang et al. reported success on the development of a highly efficient microfluidic sample preconcentration device based on an electrokinetic trapping mechanism [105].
- Riehn et al. performed a restriction mapping of DNA molecules using restriction endonucleases in nanochannels [106].
- Riehn and Austin described a new method for wetting microfluidic devices with water [107].
- Michael Chen performed a patent review of novel nanostructured devices, nanofabrication methods, and applications in nanofluidics [108].
- Park and Jung reported on the carbon nanofluidics of rapid water transport for energy applications [109].
- Zhang and Chen reported on nanofluidics for giant power harvesting [110].
- Yong and Zhang reported on thermostats and thermostat strategies for molecular dynamics simulations of nanofluidics [111].
- Wang and Lee reported on nanofluidics-based cell electroporation [112].

6.4.4 Important Computational Tools for Microflow and Nanoflow

There are several computational tools for microflow and nanoflow. Some important computation tools are listed as follows:

1. *RchyOptimyx*: RchyOptimyx is a cellular hierarchy optimization for flow cytometry [113].
2. *QUAliFiER*: QUAliFiER is an automated pipeline for quality assessment of gated flow cytometry data [114].

3. *CytoSPADE*: CytoSPADE is a high-performance analysis and visualization of high-dimensional cytometry data [115].

4. *flowPeaks*: flowPeaks is a fast unsupervised clustering for flow cytometry data via K-means and density peak finding [116].

5. *Phaedra*: Phaedra is a computational protocol-driven system for analysis and validation of high-content imaging and flow cytometry [117].

6. *FIND*: FIND is a new software tool and development platform for enhanced multicolor flow analysis [118].

7. *CytoSys*: CytoSys is a computational tool for extracting cell cycle–related expression dynamics from static data [119].

8. *SPICE*: SPICE is a computational exploration and analysis of postcytometric complex multivariate datasets [120].

9. *flowPhyto*: flowPhyto is an automated analysis of microscopic algae from continuous flow cytometric data [121].

10. *FlowFP*: FlowFP is a computational bioconductor package for fingerprinting flow cytometric data [122].

11. *FuGEFlow*: FuGEFlow is a computational data model and markup language for flow cytometry [123].

12. *flowClust*: flowClust is a bioconductor package for automated gating of flow cytometry data [124].

13. *flowCore*: flowCore is a bioconductor package for high-throughput flow cytometry [125].

6.5 Common Applications of Computational Nanoflow in General Medical Practice

As noted earlier, computational application can be useful in biomedical flow analysis. The following is a summary on how to use computational applications in medical flow analysis.

1. Application for clarification

   A good example of computational application for clarification is the usage in modeling relating to the medical flow phenomenon. Image processing and image reconstruction for the flow within a vessel can be the best example. Several published reports have already been shown in this chapter.

2. Application for prediction

   A good example of application of computational technology in the field of prediction is the simulation of the medical flow phenomenon.

Prediction might be done with the help of a specific computational technique such as computational finite difference time domain simulation.

In summary, it can be seen that computational methods can be applicable for both diagnosis and therapy in medicine:

a. In diagnosis

Computational domain method in diagnostic medical flow analysis is the main work at present. The aim is usually to answer questions on the characteristics of flowing particles in biofluids (especially in blood) as previously discussed. It can also be applied to routine medical technology or laboratory medicine work. It can help make and adjust images. The construction of medical imaging of flowing small elements in blood and cytogram of blood components are good examples.

b. In treatment

Computational techniques can be used in medical flow analysis for treatment. It is not a direct activity but it is an application. A good example is the use for predicting blood flow for cancer therapy in clinical oncology. Manipulation of medical images and reconstruction can be additionally done for planning proper cancer treatment.

There are several available programs that will help manipulation in medical flow analysis. However, MATLAB® is a good basic computational program that is widely used by many practitioner and researchers. The following examples demonstrate how MATLAB can be used.

1. Creating a graphical model

MATLAB can be used for creating a graphical model. For example, MATLAB can create a model of an isoelectric zone because of the laminar flow of electrolyte through the nanopores on the cell membrane. The code in the MATLAB Command Window is written as follows:

```
>> [x,y,z] = peaks;
>> pcolor (x,y,z);
>> shading inerp
>> hold on
>> contour (x,y,z,30,'k');
>> hold off
>>
```

2. Solving the problem of a functional equation

Considering the nature of medical flow, it will have a specific flow velocity. The flow velocity is the parameter that can be measured by several biomedical tools, and also with imaging. However, in the real clinical situation, a more important parameter to consider in the pathogenesis of a disease is acceleration. For example, MATLAB can be used for solving the acceleration function due to abnormal flow within a pathological placenta vessel. The specific code is written as follows:

```
>> y = dsolve('Dy + 5*y = 9*exp(2*t)', 'y(0) = 4');
y =
9/7*exp(2*t) + 19/7*exp(-5*t)
>> ezplot(y)
>>
```

6.5.1 Examples of Medical Research Based on Computational Nanoflow Application

Computational application in medical flow analysis in general medicine is a very interesting topic. There are many new applied techniques that lead to the advent of medical instrumentation technology.

6.5.1.1 Example 1

Finding insertional loss of ultrasonographic energy in the measurement of placental blood flow

Basically, ultrasonography is a widely used safe technique for imaging and measuring in fetomaternal medicine. As it is a radiation-free approach, it is widely used in all obstetrical clinics at present. Sound wave is mainly used and a well-known tool is the Doppler ultrasonography. In obstetrics, the Doppler ultrasonography is generally used for monitoring fetal well-being [126]. It can also be used for hearing the fetal heart [126,127]. To perform the test, Doppler ultrasonography is applied to the maternal abdomen. The ultrasonography probe has to be in contact with the abdominal wall in order to allow the ultrasound to enter into the abdominal cavity and reach the focused organ, the uterus.

On its way, there are many barriers that can absorb the sound wave. The fat within the maternal abdominal wall is one [128]. This can cause considerable insertional loss. In such a case, computational technology can find the insertional loss of ultrasonographic energy because of the maternal abdominal wall fat absorption. Theoretically, loss is usually measured at the site to be investigated, the maternal abdominal. In principle, the insertional loss depends on wavelength and frequency (time function). The basic algorithm is *concentration* = $k \times$ *wavelength* $\times$ *frequency*. As this deals with wave and time function, the computational finite difference time method can be used. An example of a relationship between change and final loss is described in Table 6.3. In this model, the variation of wavelength is the tested parameter, the observed loss is the observed parameter, and time is the controlled parameter.

TABLE 6.3

Doppler Ultrasonography
Wavelength and Insertional Loss

Wavelength (mm)	Loss (J/cm²)
0.6	1.3
0.8	1.1
1.0	0.8
1.3	1.2
1.7	1.7

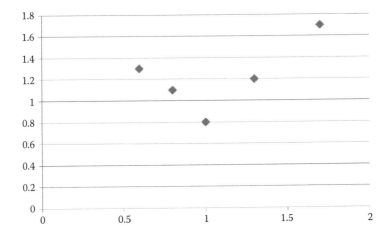

6.5.1.2 Example 2

Using standard flow cytometry to investigate the blood cell in a case of febrile illness

As already mentioned, flow cytometry is widely used in medicine as a diagnostic tool. An important application is in testing blood. In medicine, blood is routinely tested for finding disorders. When there is a pathological condition within the human body, changes in blood can be expected. There are both cells and fluids in blood. The cells and elementary components flow within the blood stream. Determining blood cells is an actual topic in medical flow analysis. The focus in the study of blood cells in medical practice includes red blood cells (erythrocytes) and white blood cells (leukocytes). Red blood cells play an important role in oxygenation. Erythrocytes contain many hemoglobin molecules, which are important oxygen carriers. White blood cells act as soldiers in human beings. They react to alien foreign bodies, including pathogens and germs. Abnormalities of both red and white blood cells can occur.

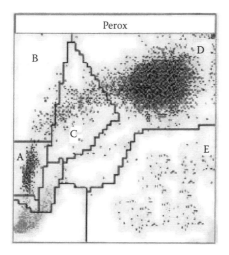

FIGURE 6.1
Complete blood count cytogram generated by automated hematology analyzer flow cytometer.

To detect an abnormality in blood cells, there is a specific medical examination, namely, a complete blood count. With advanced technology, a complete blood count can be currently done with the use of automated flow cytometry [129,130]. This method can detect the shape and number of blood cells. As these parameters are usually described in extremely small scale, there is no doubt that this is the field of nanomedicine. An example of a blood sample of a case of prolonged febrile illness examined by flow cytometry is shown in Figure 6.1. The figure shows the computer-generated diagram of a complete blood count cytogram, which is useful for a differential diagnosis of pathology in different diseases. In this example, it shows an increased population of white blood cells in the area where most of the cells are highly stained with peroxidase. These are the warning signs of possible malignancy, or leukemia.

6.5.1.3 Example 3

Using standard flow cytometry to investigate platelets of a case with bleeding abnormality

As shown in the previous example, flow cytometry is a useful tool for diagnosis in medicine. It can be used for performing blood tests. As noted, in medicine, the widely tested elements in blood are blood cells. However, not only blood cells but also other elements in the blood can be tested. The other commonly tested parameter is platelets. A platelet is the remnant of a precursor cell, namely, megakaryocyte. Platelets play an important role

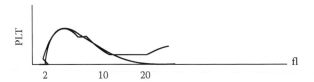

FIGURE 6.2
Complete blood count platelet scattogram generated by automated hematology analyzer flow cytometer.

in coagulation hemostasis. A person with abnormal platelets will have a problem of thrombohemostasis where bleeding is the most common clinical feature.

To detect an abnormality of platelets, a medical examination, namely, a complete blood count, can also be done. As noted, a complete blood count can be performed in a laboratory by automated flow cytometry [129,130]. With the use of flow cytometry, the shape and number of platelets can be detected, similar to blood cells [131,132]. Moreover, as the shape and number of blood cells are usually described in extremely small scale, this study is also within the scope of nanomedicine. An example of a blood sample of a case with bleeding tendency examined by flow cytometry is shown in Figure 6.2. The figure shows the computer-generated diagram, namely, a complete blood count platelet scattogram, which can be useful for a differential diagnosis of platelet abnormality.

6.5.1.4 Example 4

Using a simulation tool to investigate the fluid dynamics within a nanoneedle during cell injection

To manipulate cellular-level processes, the nanotechnique requires a very delicate tool [133,134]. A nanoneedle can be used for the process of harvesting a cellular content or infection in the cells (such as DNA). Similar to macroinjection by a standard syringe, the flow of the injection is very important. In the case of a macroinjection, the diameter of a needle and the force applied are important to determine whether the flow will be laminar or turbulent. In general, a turbulent flow is not desired as it can affect the injected site as well as the component that is injected during the process.

To assess the process of injection, an in vivo or in vitro process is generally very difficult. The use of nanoinformatics for this assessment should therefore be considered. The process of injection can be simulated. An example of the simulation of fluid dynamics within a nanoneedle during cell injection is shown. The primary assumptions include steps per second equal to 7, flow speed = 0.12, viscosity = 0.08, and small circular

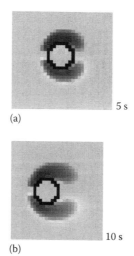

(a)

5 s

(b)

10 s

FIGURE 6.3
(See color insert.) Simulation results from fluid dynamics simulation. (a) 5 s and (b) 10 s.

barrier shape. The simulation results at 5 s and 10 s are shown in Figure 6.3. It can be seen that the expansion and diffusion of infected particles increase with time.

6.6 Conclusion

Computational technology can be widely used in several applied sciences, including medicine. With the use of the computational approach, complex medical problems can be simply manipulated and solved. It can be also used in the manipulation of medical physics problems. Among several applications in medicine, the usage of computational techniques to support the work on medical flow is very interesting. Based on the computational approach, several new medical apparatuses have been produced and launched for determining and manipulating medical fluid. In the present day, *flow* can be applied in several medical apparatuses, especially as a diagnostic tool. The medical technology of medical flow at present turns toward a more advance phase, shifting into the nanolevel. Novel techniques based on nanotechnology are also introduced as tools for manipulating medical nanoflow. Nanofluidics is the specific advent worth mentioning in this field. This technique is currently used worldwide and promises to become a standard advanced medical technology in the future.

References

1. Gehlenborg N, O'Donoghue SI, Baliga NS, Goesmann A, Hibbs MA, Kitano H, Kohlbacher O et al. Visualization of omics data for systems biology. *Nat Methods* 2010;7(3 Suppl.):S56–S68.
2. Haarala R, Porkka K. Theoddomesandomics. *Duodecim* 2002;118(11):1193–1195.
3. Haddish-Berhane N, Rickus JL, Haghighi K. The role of multiscale computational approaches for rational design of conventional and nanoparticle oral drug delivery systems. *Int J Nanomed* 2007;2(3):315–331.
4. Saliner AG, Poater A, Worth AP. Toward in silico approaches for investigating the activity of nanoparticles in therapeutic development. *IDrugs* 2008;11(10):728–732.
5. Behari J. Principles of nanoscience: An overview. *Indian J Exp Biol* 2010;48(10):1008–1019.
6. Balankin AS, Elizarraraz BE. Map of fluid flow in fractal porous medium into fractal continuum flow. *Phys Rev E Stat Nonlin Soft Matter Phys* 2012;85(5 Pt. 2): 056314.
7. Sbragaglia M, Sugiyama K. Volumetric formulation for a class of kinetic models with energy conservation. *Phys Rev E Stat Nonlin Soft Matter Phys* 2010;82(4 Pt. 2):04670.
8. Soltani M, Chen P. Numerical modeling of fluid flow in solid tumors. *PLoS ONE* 2011;6(6):e20344.
9. Guédra M, Astafyeva K, Conoir JM, Coulouvrat F, Taulier N, Thomas JL, Urbach W, Valier-Brasier T. Ultrasonic propagation in suspensions of encapsulated compressible nanoparticles. *J Acoust Soc Am* 2013;134(5):4049.
10. Cuellar JL, Llarena I, Iturri JJ, Donath E, Moya SE. A novel approach for measuring the intrinsic nanoscale thickness of polymer brushes by means of atomic force microscopy: Application of a compressible fluid model. *Nanoscale* 2013;5(23):11679–11685.
11. Tatsumi R, Yamamoto R. Velocity relaxation of a particle in a confined compressible fluid. *J Chem Phys* 2013;138(18):184905.
12. Brenner H. Proposal of a critical test of the Navier–Stokes–Fourier paradigm for compressible fluid continua. *Phys Rev E Stat Nonlin Soft Matter Phys* 2013;87(1):013014.
13. Gross M, Varnik F. Critical dynamics of an isothermal compressible nonideal fluid. *Phys Rev E Stat Nonlin Soft Matter Phys* 2012;86(6 Pt. 1):061119.
14. Godin OA. Incompressible wave motion of compressible fluids. *Phys Rev Lett* 2012;108(19):194501.
15. Felderhof BU. Dissipation in peristaltic pumping of a compressible viscous fluid through a planar duct or a circular tube. *Phys Rev E Stat Nonlin Soft Matter Phys* 2011;83(4 Pt. 2):046310.
16. Ebin DG. Motion of a slightly compressible fluid. *Proc Natl Acad Sci USA* 1975;72(2):539–542.
17. Franken H, Clément J, Cauberghs M, Van de Woestijne KP. Oscillating flow of a viscous compressible fluid through a rigid tube: A theoretical model. *IEEE Trans Biomed Eng* 1981;28(5):416–240.
18. Herlofson N. Magneto-hydrodynamic waves in a compressible fluid conductor. *Nature* 1950;165(4208):1020–1021.

19. Staroselsky II, Yakhot VV, Kida S, Orszag SA. Long-time, large-scale properties of a randomly stirred compressible fluid. *Phys Rev Lett* 1990;65(2):171–174.

20. Kim JH, Kim C, Lee SG, Hong TM, Choi JH. Viscosity of magnetorheological fluids using iron-silicon nanoparticles. *J Nanosci Nanotechnol* 2013; 13(9):6055–6059.

21. Xu M, Liu H, Zhao H, Li W. How to decrease the viscosity of suspension with the second fluid and nanoparticles? *Sci Rep* 2013;3:3137.

22. Eddi A, Winkels KG, Snoeijer JH. Influence of droplet geometry on the coalescence of low viscosity drops. *Phys Rev Lett* 2013;111(14):144502.

23. Hoang H, Galliero G. Local shear viscosity of strongly in homogeneous dense fluids: From the hard-sphere to the Lennard–Jones fluids. *J Phys Condens Matter* 2013;25(48):485001.

24. Ishikawa T, Yoshida R, Takahashi K, Inoue H, Kobayashi A. A study on EGG affected by viscosity and shear rate dependence of fluid and semi-solid diets. *Conf Proc IEEE Eng Med Biol Soc* 2013;2013:1326–1329.

25. Prakash J, Lavrenteva OM, Nir A. Interaction of bubbles in an inviscid and low-viscosity shear flow. *Phys Rev E Stat Nonlin Soft Matter Phys* 2013;88(2):023021.

26. Jestrović I, Dudik JM, Luan B, Coyle JL, Sejdić E. The effects of increased fluid viscosity on swallowing sounds in healthy adults. *Biomed Eng Online* 2013;12(1):90.

27. Holsworth RE Jr, Cho YI, Weidman J. Effect of hydration on whole blood viscosity in firefighters. *Altern Ther Health Med* 2013;19(4):44–49.

28. Gachelin J, Miño G, Berthet H, Lindner A, Rousselet A, Clément E. Non-Newtonian viscosity of *Escherichia coli* suspensions. *Phys Rev Lett* 2013; 110(26):268103.

29. Llovell F, Marcos RM, Vega LF. Free-volume theory coupled with soft-SAFT for viscosity calculations: Comparison with molecular simulation and experimental data. *J Phys Chem B* 2013;117(27):8159–8171.

30. Cheng L, Li Y, Grosh K. Including fluid shear viscosity in a structural acoustic finite element model using a scalar fluid representation. *J Comput Phys* 2013;247:248–261.

31. Zhang W, Qian Y, Lin J, Lv P, Karunanithi K, Zeng M. Hemodynamic analysis of renal artery stenosis using computational fluid dynamics technology based on unenhanced steady-state free precession magnetic resonance angiography: Preliminary results. *Int J Cardiovasc Imag* February 2014;30(2):367–375. [Epub ahead of print].

32. Throckmorton AL, Tahir SA, Lopes SP, Rangus OM, Sciolino MG. Steady and transient flow analysis of a magnetically levitated pediatric VAD: Time varying boundary conditions. *Int J Artif Organs* 2013;36(10):693–699.

33. Ghatage D, Chatterji A. Modeling steady-state dynamics of macromolecules in exponential-stretching flow using multiscale molecular-dynamics-multiparticle-collision simulations. *Phys Rev E Stat Nonlin Soft Matter Phys* 2013;88(4 Pt. 1): 043303.

34. Martinez C, Henao A, Rodriguez JE, Padgett KR, Ramaswamy S. Monitoring steady flow effects on cell distribution in engineered valve tissues by magnetic resonance imaging. *Mol Imag* 2013;12(7):470–482.

35. Shehzad SA, Alsaedi A, Hayat T. Hydromagnetic steady flow of Maxwell fluid over a bidirectional stretching surface with prescribed surface temperature and prescribed surface heat flux. *PLoS ONE* 2013;8(7):e68139.

36. Larsen PS, Riisgård HU. Validation of the flow-through chamber (FTC) and steady-state (SS) methods for clearance rate measurements in bivalves. *Biol Open* 2012;1(1):6–11.

37. Bedkihal S, Kumaradas JC, Rohlf K. Steady flow through a constricted cylinder by multiparticle collision dynamics. *Biomech Model Mechanobiol* 2013;12(5):929–939.

38. Potvin G. Hydrodynamics of the turbulent point-spread function. *J Opt Soc Am A Opt Image Sci Vis* 2013;30(7):1342–1349.

39. Pravin S, Reidenbach MA. Simultaneous sampling of flow and odorants by crustacean scan aid searches within a turbulent plume. *Sensors (Basel)* 2013;13(12):16591–16610.

40. Bourgogne E, Soichot M, Latour C, Laprévote O. Rugged and accurate quantitation of colchicines in human plasma to support colchicines poisoning monitoring by using turbulent-flow LC-MS/MS analysis. *Bioanalysis* 2013;5(23):2889–2896.

41. Kwon Y, Lee K, Park M, Koo K, Lee J, Doh Y, Lee S, Kim D, Jung Y. Temperature dependence of convective heat transfer with Al_2O_3 nanofluids in the turbulent flow region. *J Nanosci Nanotechnol* 2013;13(12):7902–7905.

42. Skartlien R, Sollum E, Schumann H. Droplet size distributions in turbulent emulsions: Breakup criteria and surfactant effects from direct numerical simulations. *J Chem Phys* 2013;139(17):174901.

43. Yadav AS, Bhagoria JL. Modeling and simulation of turbulent flows through a solar air heater having square-sectioned transverse rib roughness on the absorber plate. *Sci World J* 2013;2013:827131.

44. Shetty DA, Frankel SH. Assessment of stretched vortex subgrid-scale models for LES of incompressible in homogeneous turbulent flow. *Int J Numer Methods Fluids* 2013;73(2). doi: 10.1002/fld.3793

45. Wu P, Wan M, Matthaeus WH, Shay MA, Swisdak M. Von Kármán energy decay and heating of protons and electrons in a kinetic turbulent plasma. *Phys Rev Lett* 2013;111(12):121105.

46. Kim K, Sureshkumar R. Spatio temporal evolution of hairpin eddies, Reynolds stress, and polymer torque in polymer drag-reduced turbulent channel flows. *Phys Rev E Stat Nonlin Soft Matter Phys* 2013;87(6):063002.

47. Nootz G, Hou W, Dalgleish FR, Rhodes WT. Determination of flow orientation of an optically active turbulent field by means of a single beam. *Opt Lett* 2013;38(13):2185–2187.

48. Lee BK, Mohan BR, Byeon SH, Lim KS, Hong EP. Evaluating the performance of a turbulent wet scrubber for scrubbing particulate matter. *J Air Waste Manag Assoc* 2013;63(5):499–506.

49. Klinkenberg J, Sardina G, de Lange HC, Brandt L. Numerical study of laminar-turbulent transition in particle-laden channel flow. *Phys Rev E Stat Nonlin Soft Matter Phys* 2013;87(4):043011.

50. Sahore V, Fritsch I. Flat flow profiles achieved with microfluidics generated by redox-magnetohydrodynamics. *Anal Chem* 2013;85(24):11809–11816.

51. Khan I, Ali F, Shafie S. Stokes' second problem for magnetohydrodynamics flow in a Burgers' fluid: The cases $\gamma=\lambda^2/4$ and $\gamma>\lambda^2/4$. *PLoS ONE* 2013;8(5):e61531.

52. Arter W. Potential vorticity formulation of compressible magnetohydrodynamics. *Phys Rev Lett* 2013;110(1):015004.

53. Gao F, Kreidermacher A, Fritsch I, Heyes CD. 3D imaging of flow patterns in an internally-pumped microfluidic device: Redox magnetohydrodynamics and electrochemically-generated density gradients. *Anal Chem* 2013;85(9):4414–4422.
54. Morales JA, Bos WJ, Schneider K, Montgomery DC. Intrinsic rotation of toroidally confined magnetohydrodynamics. *Phys Rev Lett* 2012;109(17):175002.
55. Ensafi AA, Nazari Z, Fritsch I. Redox magnetohydrodynamics enhancement of stripping voltammetry of lead(II), cadmium(II) and zinc(II) ions using 1,4-benzoquinone as an alternative pumping species. *Analyst* 2012;137(2):424–431.
56. Krstulovic G, Brachet ME, Pouquet A. Alfvén waves and ideal two-dimensional Galerkin truncated magnetohydrodynamics. *Phys Rev E Stat Nonlin Soft Matter Phys* 2011;84(1 Pt. 2):016410.
57. Lyutikov M, Hadden S. Relativistic magnetohydrodynamics in one dimension. *Phys Rev E Stat Nonlin Soft Matter Phys* 2012;85(2 Pt. 2):026401.
58. Cheah LT, Fritsch I, Haswell SJ, Greenman J. Evaluation of heart tissue viability under redox-magnetohydrodynamics conditions: Toward fine-tuning flow in biological microfluidics applications. *Biotechnol Bioeng* 2012;109(7):1827–1834. doi:10.1002/bit.24426.
59. Onofri M, Malara F, Veltri P. Effects of an isotropic thermal conductivity in magnetohydrodynamics simulations of a reversed-field pinch. *Phys Rev Lett* 2010;105(21):215006.
60. Benzi R, Pinton JF. Magnetic reversals in a simple model of magnetohydrodynamics. *Phys Rev Lett* 2010;105(2):024501.
61. Nigro G, Carbone V. Magnetic reversals in a modified shell model for magnetohydrodynamics turbulence. *Phys Rev E Stat Nonlin Soft Matter Phys* 2010;82(1 Pt. 2):016313.
62. Chatterjee D, Amiroudine S. Lattice kinetic simulation of nonisothermal magnetohydrodynamics. *Phys Rev E Stat Nonlin Soft Matter Phys* 2010;81(6 Pt. 2):066703.
63. Mininni PD, Pouquet A. Finite dissipation and intermittency in magnetohydrodynamics. *Phys Rev E Stat Nonlin Soft Matter Phys* 2009;80(2 Pt. 2):025401.
64. Lessinnes T, Carati D, Verma MK. Energy transfers in shell models for magnetohydrodynamics turbulence. *Phys Rev E Stat Nonlin Soft Matter Phys* 2009;79(6 Pt. 2):066307.
65. Lee E, Brachet ME, Pouquet A, Mininni PD, Rosenberg D. Paradigmatic flow for small-scale magnetohydrodynamics: Properties of the ideal case and the collision of current sheets. *Phys Rev E Stat Nonlin Soft Matter Phys* 2008;78(6 Pt. 2):066401.
66. Ohkitani K, Constantin P. Two-and three-dimensional magnetic reconnection observed in the Eulerian–Lagrangian analysis of magnetohydrodynamics equations. *Phys Rev E Stat Nonlin Soft Matter Phys* 2008;78(6 Pt. 2):066315.
67. Servidio S, Matthaeus WH, Carbone V. Ergodicity of ideal Galerkin three-dimensional magnetohydrodynamics and Hall magnetohydrodynamics models. *Phys Rev E Stat Nonlin Soft Matter Phys* 2008;78(4 Pt. 2):046302.
68. Lu T, Du J, Samulyak R. A numerical algorithm for magnetohydrodynamics of ablated materials. *J Nanosci Nanotechnol* 2008;8(7):3674–3685.
69. Simakov AN, Chacón L. Quantitative, comprehensive, analytical model for magnetic reconnection in Hall magnetohydrodynamics. *Phys Rev Lett* 2008;101(10):105003.

70. Chacón L, Simakov AN, Zocco A. Steady-state properties of driven magnetic reconnection in 2D electron magnetohydrodynamics. *Phys Rev Lett* 2007;99(23):235001.

71. Kim EJ. Consistent theory of turbulent transport in two-dimensional magnetohydrodynamics. *Phys Rev Lett* 2006;96(8):084504.

72. De Moortel I. Propagating magnetohydrodynamics waves in coronal loops. *Philos Trans A Math Phys Eng Sci* 2006;364(1839):461–471.

73. Wheatley V, Pullin DI, Samtaney R. Stability of an impulsively accelerated density interface in magnetohydrodynamics. *Phys Rev Lett* 2005;95(12):125002.

74. Graham JP, Mininni PD, Pouquet A. Cancellation exponent and multifractal structure in two-dimensional magnetohydrodynamics: Direct numerical simulations and Lagrangian averaged modeling. *Phys Rev E Stat Nonlin Soft Matter Phys* 2005;72(4 Pt. 2):045301.

75. Hatami M, Hatami J, Ganji DD. Computer simulation of MHD blood conveying gold nanoparticles as a third grade non-Newtonian nanofluid in a hollow porous vessel. *Comput Methods Prog Biomed* February 2014;113(2):632–641.

76. Voronov RS, Stalker TJ, Brass LF, Diamond SL. Simulation of intrathrombus fluid and solute transport using in vivo clot structures with single platelet resolution. *Ann Biomed Eng* 2013;41(6):1297–1307.

77. Dong J, Wong KK, Tu J. Hemodynamics analysis of patient-specific carotid bifurcation: A CFD model of downstream peripheral vascular impedance. *Int J Numer Method Biomed Eng* 2013;29(4):476–491.

78. Omodaka S, Sugiyama S, Inoue T, Funamoto K, Fujimura M, Shimizu H, Hayase T, Takahashi A, Tominaga T. Local hemodynamics at the rupture point of cerebral aneurysms determined by computational fluid dynamics analysis. *Cerebrovasc Dis* 2012;34(2):121–129.

79. Naito T, Miyachi S, Matsubara N, Isoda H, Izumi T, Haraguchi K, Takahashi I, Ishii K, Wakabayashi T. Magnetic resonance fluid dynamics for intracranial aneurysms—Comparison with computed fluid dynamics. *Acta Neurochir (Wien)* 2012;154(6):993–1001.

80. Qian Y, Takao H, Umezu M, Murayama Y. Risk analysis of unruptured aneurysms using computational fluid dynamics technology: Preliminary results. *AJNR Am J Neuroradiol* 2011;32(10):1948–1955.

81. Fukuda I, Osanai S, Shirota M, Inamura T, Yanaoka H, Minakawa M, Fukui K. Computer-simulated fluid dynamics of arterial perfusion in extracorporeal circulation: From reality to virtual simulation. *Int J Artif Organs* 2009;32(6):362–370.

82. Wood NB, Weston SJ, Kilner PJ, Gosman AD, Firmin DN. Combined MR imaging and CFD simulation of flow in the human descending aorta. *J Magn Reson Imag* 2001;13(5):699–713.

83. Lihavainen E, Mäkelä J, Spelbrink JN, Ribeiro AS. Mytoe: Automatic analysis of mitochondrial dynamics. *Bioinformatics* 2012;28(7):1050–1051.

84. Lysholm F, Andersson B, Persson B. FAAST: Flow-space assisted alignment search tool. *BMC Bioinformatics* 2011;12:293.

85. Ribalet F, Schruth DM, Armbrust EV. flowPhyto: Enabling automated analysis of microscopic algae from continuous flowcytometric data. *Bioinformatics* 2011;27(5):732–733.

86. Lee K, Hahne F, Sarkar D, Gentleman R. iFlow: A graphical user interface for flow cytometry tools in bioconductor. *Adv Bioinformatics* 2009;2009:103839.

87. Stojmirović A, Yu YK. ITMProbe: Analyzing information flow in protein networks. *Bioinformatics* 2009;25(18):2447–2449.
88. Muzic RF Jr, Cornelius S. COMKAT: Compartment model kinetic analysis tool. *J Nucl Med* 2001;42(4):636–645.
89. Stojmirović A, Bliskovsky A, Yu YK. CytoITMprobe: A network information flow plugin for Cytoscape. *BMC Res Notes* 2012;5:237.
90. Fodinger M, Sunder-Plassmann G, Wagner OF. Trends in molecular diagnosis. *Wien Klin Wochenschr* 1999;111:315–319.
91. Besselink GA, Vulto P, Lammertink RG, Schlautmann S, vanden Berg A, Olthuis W, Engbers GH, Schasfoort RB. Electroosmotic guiding of sample flows in a laminar flow chamber. *Electrophoresis* 2004;25:3705–3711.
92. Soo Ko J, Yoon HC, Yang H, Pyo HB, Hyo Chung K, Jin Kim S, Tae Kim Y. A polymer-based microfluidic device for immunosensing biochips. *Lab Chip* 2003;3:106–113.
93. Ionescu RE, Marks RS, Gheber LA. Manufacturing of nanochannels with controlled dimensions using protease nanolithography. *Nano Lett* 2005;5:821–827.
94. Mijatovic D, Eijkel JC, vanden Berg A. Technologies for nanofluidic systems: Top-down vs. bottom-up—A review. *Lab Chip* 2005;5(5):492–500.
95. Lai S, Wang S, Luo J, Lee LJ, Yang ST, Madou MJ. Design of a compact disk-like microfluidic platform for enzyme-linked immunosorbent assay. *Anal Chem* 2004;76(7):1832–1837.
96. Morier P, Vollet C, Michel PE, Reymond F, Rossier JS. Gravity-induced convective flow in microfluidic systems: Electrochemical characterization and application to enzyme-linked immunosorbent assay tests. *Electrophoresis* 2004;25(21–22):3761–3768.
97. Herrmann M, Veres T, Tabrizian M. Enzymatically-generated fluorescent detection in micro-channels with internal magnetic mixing for the development of parallel microfluidic ELISA. *Lab Chip* 2006;6(4):555–560.
98. Eteshola E, Balberg M. Microfluidic ELISA: On-chip fluorescence imaging. *Biomed Microdevices* 2004;6(1):7–9.
99. Ding X, Li P, Lin SC, Stratton ZS, Nama N, Guo F, Slotcavage D, Mao X, Shi J, Costanzo F, Huang TJ. Surface acoustic wave microfluidics. *Lab Chip* 2013;13(18):3626–3649.
100. Jin BJ, Ko EA, Namkung W, Verkman AS. Microfluidics platform for single-shot dose–response analysis of chloride channel-modulating compounds. *Lab Chip* 2013;13(19):3862–3867.
101. Liu Y, Tian M, Zhang H. Microfluidics for synthesis of peptide-based PET tracers. *Biomed Res Int* 2013;2013:839683.
102. Colace TV, Tormoen GW, McCarty OJ, Diamond SL. Microfluidics and coagulation biology. *Annu Rev Biomed Eng* 2013;15:283–303.
103. Al-Hetlani E. Forensic drug analysis and microfluidics. *Electrophoresis* 2013;34(9–10):1262–1272.
104. Parikesit GO, Markesteijn AP, Kutchoukov VG, Piciu O, Bossche A, Westerweel J, Garini Y, Young IT. Electroosmotic flow analysis of a branched U-turn nanofluidic device. *Lab Chip* 2005;5(10):1067–1074.
105. Wang YC, Stevens AL, Han J. Million-fold preconcentration of proteins and peptides by nanofluidic filter. *Anal Chem* 2005;77(14):4293–4299.

106. Riehn R, Lu M, Wang YM, Lim SF, Cox EC, Austin RH. Restriction mapping in nanofluidic devices. *Proc Natl Acad Sci USA* 2005;102(29):10012–10016.

107. Riehn R, Austin RH. Wetting micro- and nanofluidic devices using supercritical water. *Anal Chem* 2006;78(16):5933–5934.

108. Michael Chen CY. Patent review of novel nanostructured devices, nanofabrication methods and applications in nanofluidics and nanomedicine. *Recent Pat Nanotechnol* 2012;6(2):114–123.

109. Park HG, Jung Y. Carbon nanofluidics of rapid water transport for energy applications. *Chem Soc Rev* January 2014;43(2):565–576. [Epub ahead of print].

110. Zhang L, Chen X. Nanofluidics for giant power harvesting. *Angew Chem Int Ed Engl* 2013;52(30):7640–7641.

111. Yong X, Zhang LT. Thermostats and thermostat strategies for molecular dynamics simulations of nanofluidics. *J Chem Phys* 2013;138(8):084503.

112. Wang S, Lee LJ. Micro-/nanofluidics based cell electroporation. *Biomicrofluidics* 2013;7(1):11301.

113. Aghaeepour N, Jalali A, O'Neill K, Chattopadhyay PK, Roederer M, Hoos HH, Brinkman RR. RchyOptimyx: Cellular hierarchy optimization for flowcytometry. *Cytometry A* 2012;81(12):1022–1030.

114. Finak G, Jiang W, Pardo J, Asare A, Gottardo R. QUALiFiER: An automated pipeline for quality assessment of gated flowcytometry data. *BMC Bioinformatics* 2012;13:252.

115. Linderman MD, Bjornson Z, Simonds EF, Qiu P, Bruggner RV, Sheode K, Meng TH, Plevritis SK, Nolan GP. CytoSPADE: High-performance analysis and visualization of high-dimensional cytometry data. *Bioinformatics* 2012;28(18):2400–2401.

116. Ge Y, Sealfon SC. FlowPeaks: A fast unsupervised clustering for flowcytometry data via K-means and density peak finding. *Bioinformatics* 2012;28(15):2052–2058.

117. Cornelissen F, Cik M, Gustin E. Phaedra, a protocol-driven system for analysis and validation of high-content imaging and flowcytometry. *J Biomol Screen* 2012;17(4):496–506.

118. Dabdoub SM, Ray WC, Justice SS. FIND: A new software tool and development platform for enhanced multicolor flowanalysis. *BMC Bioinformatics* 2011;12:145.

119. Avva J, Weis MC, Soebiyanto RP, Jacobberger JW, Sreenath SN. CytoSys: A tool for extracting cell-cycle-related expression dynamics from static data. *Methods Mol Biol* 2011;717:171–193.

120. Roederer M, Nozzi JL, Nason MC. SPICE: Exploration and analysis of post-cytometric complex multivariate datasets. *Cytometry A* 2011;79(2):167–174.

121. Ribalet F, Schruth DM, Armbrust EV. flowPhyto: Enabling automated analysis of microscopic algae from continuous flow cytometric data. *Bioinformatics* 2011;27(5):732–733.

122. Rogers WT, Holyst HA. FlowFP: A bioconductor package for fingerprinting flow cytometric data. *Adv Bioinformatics* 2009;2009:193947.

123. Qian Y, Tchuvatkina O, Spidlen J, Wilkinson P, Gasparetto M, Jones AR, Manion FJ, Scheuermann RH, Sekaly RP, Brinkman RR. FuGEFlow: Datamodel and markup language for flowcytometry. *BMC Bioinformatics* 2009;10:184.

124. Lo K, Hahne F, Brinkman RR, Gottardo R. flowClust: A bioconductor package for automated gating of flowcytometry data. *BMC Bioinformatics* 2009;10:145.

125. Hahne F, LeMeur N, Brinkman RR, Ellis B, Haaland P, Sarkar D, Spidlen J, Strain E, Gentleman R. flowCore: A Bioconductor package for high throughput flowcytometry. *BMC Bioinformatics* 2009;10:106.
126. Campbell S, Bewley S, Cohen-Overbeek T. Investigation of the uteroplacental circulation by Doppler ultrasound. *Semin Perinatol* 1987;11(4):362–368.
127. Reed KL. Fetal and neonatal cardiac assessment with Doppler. *Semin Perinatol* 1987;11(4):347–356.
128. Nedev PI, Nedeva AI, Uchikov AP. Ultrasonographic projection of a pathologic site on the anterior abdominal wall. Our own experience. *FoliaMed (Plovdiv)* 2003;45(2):17–22.
129. Kim YR, van't Oever R, Landayan M, Bearden J. Automated red blood cell differential analysis on a multi-angle lightscatter/fluorescence hematology analyzer. *Cytometry B Clin Cytom* 2003;56(1):43–54.
130. Watanabe T, Sanada H. Diagnostic problems of hematological disorder using automated blood cell analyzer. *Rinsho Byori* 1996;44(8):729–735.
131. Pati HP, Jain S. Flowcytometry in hematological disorders. *Indian J Pediatr* 2013;80(9):772–778.
132. Mininkova AI. Investigation of platelets by the flow cyto fluorometric technique (a review of literature). Part 2. *Klin Lab Diagn* 2011;4:25–30.
133. Kolhar P, Doshi N, Mitragotri S. Polymer nanoneedle-mediated intracellular drug delivery. *Small* 2011;7(14):2094–2100.
134. Obataya I, Nakamura C, Han S, Nakamura N, Miyake J. Nanoscale operation of a living cell using an atomic force microscope with a nanoneedle. *Nano Lett* 2005;5(1):27–30.

Appendix A: Material and Physical Constants

A.1 Common Material Constants

TABLE A.1

Approximate Conductivity at 20°C

Material	Conductivity (S/m)
1. Conductors	
Silver	6.3×10^7
Copper (standard annealed)	5.8×10^7
Gold	4.5×10^7
Aluminum	3.5×10^7
Tungsten	1.8×10^7
Zinc	1.7×10^7
Brass	1.1×10^7
Iron (pure)	10^7
Lead	5×10^7
Mercury	10^6
Carbon	3×10^7
Water (sea)	4.8
2. Semiconductors	
Germanium (pure)	2.2
Silicon (pure)	4.4×10^{-4}
3. Insulators	
Water (distilled)	10^{-4}
Earth (dry)	10^{-5}
Bakelite	10^{-10}
Paper	10^{-11}
Glass	10^{-12}
Porcelain	10^{-12}
Mica	10^{-15}
Paraffin	10^{-15}
Rubber (hard)	10^{-15}
Quartz (fused)	10^{-17}
Wax	10^{-17}

TABLE A.2

Approximate Dielectric Constant and Dielectric Strength

Material	Dielectric Constant (or Relative Permittivity) (Dimensionless)	Strength, E (V/m)
Barium titanate	1200	7.5×10^6
Water (sea)	80	—
Water (distilled)	8.1	—
Nylon	8	—
Paper	7	12×10^6
Glass	5–10	35×10^6
Mica	6	70×10^6
Porcelain	6	—
Bakelite	5	20×10^6
Quartz (fused)	5	30×10^6
Rubber (hard)	3.1	25×10^6
Wood	2.5–8.0	—
Polystyrene	2.55	—
Polypropylene	2.25	—
Paraffin	2.2	30×10^6
Petroleum oil	2.1	12×10^6
Air (1 atm)	1	3×10^6

TABLE A.3

Relative Permeability

Material	Relative Permeability, μ_r
1. Diamagnetic	
Bismuth	0.999833
Mercury	0.999968
Silver	0.9999736
Lead	0.9999831
Copper	0.9999906
Water	0.9999912
Hydrogen (STP)	≈ 1.0
2. Paramagnetic	
Oxygen (STP)	0.999998
Air	1.00000037
Aluminum	1.000021
Tungsten	1.00008
Platinum	1.0003
Manganese	1.001
3. Ferromagnetic	
Cobalt	250
Nickel	600
Soft iron	5000
Silicon iron	7000

TABLE A.4

Approximate Conductivity for Biological Tissue

Material	Conductivity (S/m)	Frequency
Blood	0.7	0 (DC)
Bone	0.01	0 (DC)
Brain	0.1	10^2–10^6 Hz
Breast fat	0.2–1	0.4–5 GHz
Breast tumor	0.7–3	0.4–5 GHz
Fat	0.1–0.3	0.4–5 GHz
	0.03	10^2–10^6 Hz
Muscle	0.4	10^2–10^6 Hz
Skin	0.001	1 kHz
	0.1	1 MHz

TABLE A.5

Approximate Dielectric Constant for Biological Tissue

Material	Dielectric Constant (Relative Permittivity)	Frequency
Blood	10^5	1 kHz
Bone	3,000–10,000	0 (DC)
Brain	10^7	100 Hz
	10^3	1 MHz
Breast fat	5–50	0.4–5 GHz
Breast tumor	47–67	0.4–5 GHz
Fat	5	0.4–5 GHz
	10^6	100 Hz
	10	1 MHz
Muscle	10^6	1 kHz
	10^3	1 MHz
Skin	10^6	1 kHz
	10^3	1 MHz

A.2 Physical Constants

Quantity	Best Experimental Value	Approximate Value for Problem Work
Avogadro's number (/kg mol)	6.0228×10^{26}	6×10^{26}
Boltzmann constant (J/K)	1.38047×10^{-23}	1.38×10^{-23}
Electron charge (C)	-1.6022×10^{-19}	-1.6×10^{-19}
Electron mass (kg)	9.1066×10^{-31}	9.1×10^{-31}
Permittivity of free space (F/m)	8.854×10^{-12}	$\dfrac{10^{-9}}{36\pi}$
Permeability of free space (H/m)	$4\pi \times 10^{-7}$	12.6×10^{-7}
Intrinsic impedance of free space (O)	376.6	120π
Speed of light in free space or vacuum (m/s)	2.9979×10^{8}	3×10^{8}
Proton mass (kg)	1.67248×10^{-27}	1.67×10^{-27}
Neutron mass (kg)	1.6749×10^{-27}	1.67×10^{-27}
Planck's constant (J s)	6.6261×10^{-34}	6.62×10^{-34}
Acceleration due to gravity (m/s^2)	9.8066	9.8
Universal constant of gravitation (m^2/kg s^2)	6.658×10^{-11}	6.66×10^{-11}
Electron volt (J)	1.6030×10^{-19}	1.6×10^{-19}
Gas constant (J/mol K)	8.3145	8.3

Appendix B: Photon Equations, Index of Refraction, Electromagnetic Spectrum, and Wavelengths of Commercial Lasers

B.1 Photon Energy, Frequency, Wavelength

Photon energy (J)	Planck's constant × frequency
Photon energy (eV)	$\dfrac{\text{Planck's constant} \times \text{frequency}}{\text{Electron charge}}$
Photon energy (cm^{-1})	$\dfrac{\text{Frequency}}{\text{Speed of light in vacuum}}$
Photon frequency (Hz)	$\dfrac{1 \text{ (cycle)}}{\text{Period (s)}}$
Photon wavelength (µm)	$\dfrac{\text{Speed of light in free space}}{\text{Frequency}}$

B.2 Index of Refraction for Common Substances

Substance	Index of Refraction
Air	1.000293
Diamond	2.24
Ethyl alcohol	1.36
Fluorite	1.43
Fused quartz	1.46
Crown glass	1.52
Flint glass	1.66
Glycerin	1.47
Ice	1.31
Polystyrene	1.49
Rock salt	1.54
Water	1.33

B.3 Electromagnetic Spectrum

See Figure B.1 and Table B.1.

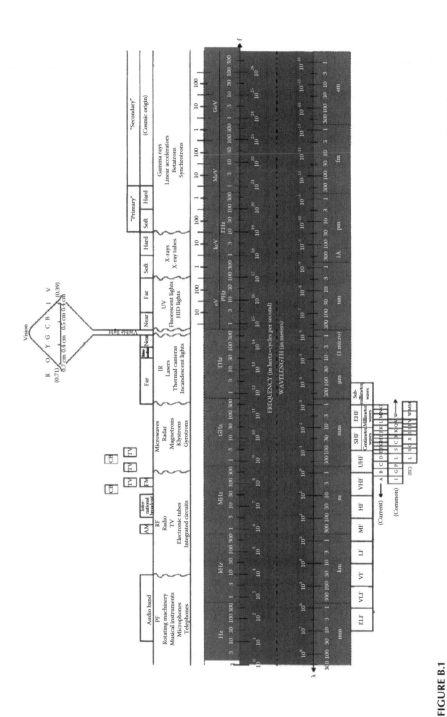

FIGURE B.1
Simplified chart of the electromagnetic spectrum. (From Whitaker, J.C., *The Electronics Handbook*, CRC Press, Boca Raton, FL, 1996.)

TABLE B.1

Approximate Common Optical Wavelength Ranges of Light

Color	Wavelength
Ultraviolet region	10–380 nm
Visible region	380–750 nm
Violet	380–450 nm
Blue	450–495 nm
Green	495–570 nm
Yellow	570–590 nm
Orange	590–620 nm
Red	620–750 nm
Infrared	750 nm–1 mm

B.4 Wavelengths of Commercially Available Lasers

See Figure B.2.

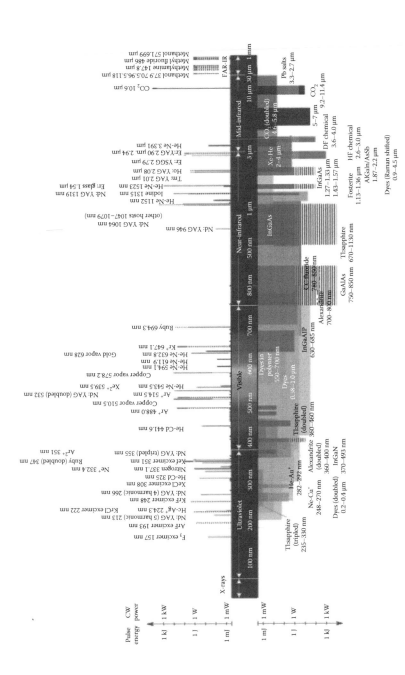

FIGURE B.2
(See color insert.) Wavelengths of commercially available lasers. (From Weber, M.J., *Handbook of Laser Wavelengths*, CRC Press, Boca Raton, FL, 1999.)

References

1. Weber, M.J., *Handbook of Laser Wavelengths*, CRC Press, Boca Raton, FL, 1999, ISBN 0-8493-3508-6.
2. Whitaker, J.C., *The Electronics Handbook*, CRC Press, Boca Raton, FL, 1996.

Appendix C: Symbols, Formulas, and Periodic Table

C.1 Greek Alphabet

Uppercase	Lowercase	Name
A	α	Alpha
B	β	Beta
Γ	γ	Gamma
Δ	δ	Delta
E	ε	Epsilon
Z	ζ	Zeta
H	H	Eta
Θ	θ, ϑ	Theta
I	ι	Iota
K	κ	Kappa
Λ	λ	Lambda
M	μ	Mu
N	ν	Nu
Ξ	ξ	Xi
O	o	Omicron
Π	π	Pi
P	ρ	Rho
Σ	σ	Sigma
T	τ	Tau
Υ	υ	Upsilon
Φ	φ, φ	Phi
X	χ	Chi
Ψ	ψ	Psi
Ω	ω	Omega

C.2 International System of Units (SI) Prefixes

Power	Prefix	Symbol	Power	Prefix	Symbol
10^{-35}	Stringo	—	10^0	—	—
10^{-24}	Yocto	y	10^1	Deca	da
10^{-21}	Zepto	z	10^2	Hecto	h
10^{-18}	Atto	A	10^3	Kilo	k
10^{-15}	Femto	f	10^6	Mega	M
10^{-12}	Pico	p	10^9	Giga	G
10^{-9}	Nano	n	10^{12}	Tera	T
10^{-6}	Micro	μ	10^{15}	Peta	P
10^{-3}	Milli	m	10^{18}	Exa	E
10^{-2}	Centi	c	10^{21}	Zetta	Z
10^{-1}	Deci	d	10^{24}	Yotta	Y

C.3 Trigonometric Identities

$$\cot\theta = \frac{1}{\tan\theta}, \quad \sec\theta = \frac{1}{\cos\theta}, \quad \operatorname{cosec}\theta = \frac{1}{\sin\theta}$$

$$\tan\theta = \frac{\sin\theta}{\cos\theta}, \quad \cot\theta = \frac{\cos\theta}{\sin\theta}$$

$$\sin^2\theta + \cos^2\theta = 1, \ \tan^2\theta + 1 = \sec^2\theta, \ \cot^2\theta + 1 = \csc^2\theta$$

$$\sin(-\theta) = -\sin\theta, = \cos(-\theta) = \cos\theta, \ \tan(-\theta) = -\tan\theta$$

$$\csc(-\theta) = -\csc\theta, \ \sec(-\theta) = \sec\theta, \ \cot(-\theta) = -\cot\theta$$

$$\cos(\theta_1 \pm \theta_2) = \cos\theta_1\cos\theta_2 \pm \sin\theta_1\sin\theta_2$$

$$\sin(\theta_1 \pm \theta_2) = \sin\theta_1\cos\theta_2 \pm \cos\theta_1\sin\theta_2$$

$$\tan(\theta_1 \pm \theta_2) = \frac{\tan\theta_1 \pm \tan\theta_2}{1 \mp \tan\theta_1 \pm \tan\theta_2}$$

$$\cos\theta_1\cos\theta_2 = \frac{1}{2}\left[\cos\left(\theta_1 + \theta_2\right) + \cos\left(\theta_1 - \theta_2\right)\right]$$

$$\sin\theta_1\sin\theta_2 = \frac{1}{2}\left[\sin\left(\theta_1 + \theta_2\right) + \sin\left(\theta_1 - \theta_2\right)\right]$$

$$\sin\theta_1\cos\theta_2 = \frac{1}{2}\left[\sin\left(\theta_1 + \theta_2\right) + \sin\left(\theta_1 - \theta_2\right)\right]$$

$$\cos\theta_1\sin\theta_2 = \frac{1}{2}\left[\sin\left(\theta_1 + \theta_2\right) + \sin\left(\theta_1 - \theta_2\right)\right]$$

$$\sin\theta_1 + \sin\theta_2 = 2\sin\left(\frac{\theta_1 + \theta_2}{2}\right)\cos\left(\frac{\theta_1 - \theta_2}{2}\right)$$

$$\sin\theta_1 - \sin\theta_2 = 2\cos\left(\frac{\theta_1 + \theta_2}{2}\right)\sin\left(\frac{\theta_1 - \theta_2}{2}\right)$$

$$\cos\theta_1 + \cos\theta_2 = 2\cos\left(\frac{\theta_1 + \theta_2}{2}\right)\cos\left(\frac{\theta_1 - \theta_2}{2}\right)$$

$$\cos\theta_1 - \cos\theta_2 = 2\sin\left(\frac{\theta_1 + \theta_2}{2}\right)\sin\left(\frac{\theta_1 - \theta_2}{2}\right)$$

$$a\sin\theta - b\cos\theta = \sqrt{a^2 + b^2}\,\cos(\theta + \phi), \quad \text{where } \phi = \tan^{-1}\left(\frac{b}{a}\right)$$

$$a\sin\theta - b\cos\theta = \sqrt{a^2 + b^2}\,\sin(\theta + \phi), \quad \text{where } \phi = \tan^{-1}\left(\frac{b}{a}\right)$$

$\cos(90° - \theta) = \sin\theta$, $\sin(90° - \theta) = \cos\theta$, $\tan(90° - \theta) = \cot\theta$

$\cos(90° - \theta) = \tan\theta$, $\sec(90° - \theta) = \text{cosec}\,\theta$, $\text{cosec}(90° - \theta) = \sec\theta$

$\cos(\theta \pm 90°) = \mp\sin\theta$, $\sec(\theta \pm 90°) = \pm\sin\theta$, $\tan(\theta \pm 90°) = -\cot\theta$

$\cos(\theta \pm 180°) = -\cos\theta$, $\sin(\theta \pm 180°) = -\sin\theta$, $\tan(\theta \pm 180°) = \tan\theta$

$\cos 2\theta = \cos^2\theta - \sin^2\theta$, $\cos 2\theta = 1 - 2\sin^2\theta$, $\cos 2\theta = 2\cos^2\theta - 1$

$\sin 2\theta = 2\sin\theta\cos\theta$, $\tan 2\theta = \dfrac{2\tan\theta}{1 - \tan^2\theta}$

$\cos 3\theta = 4\cos^3\theta - 3\sin\theta$

$\sin 3\theta = 3\sin\theta - 4\sin^3\theta$

$\sin\dfrac{\theta}{2} = \pm\sqrt{\dfrac{1 - \cos\theta}{2}}$, $\quad \cos\dfrac{\theta}{2} \pm \sqrt{\dfrac{1 + \cos\theta}{2}}$,

$\sin\theta = \dfrac{e^{j\theta} - e^{j\theta}}{2j}$, $\quad \cos\theta = \dfrac{e^{j\theta} + e^{j\theta}}{2}\left(j = \sqrt{-1}\right)$, $\quad \tan\theta = \dfrac{e^{j\theta} + e^{-j\theta}}{j\left(e^{j\theta} + e^{-j\theta}\right)}$

$e^{\pm j\theta} = \cos\theta \pm j\sin\theta$ (Euler's identity)

$1\text{ rad} = 57.296°$

$\pi = 3.1416$

C.4 Hyperbolic Functions

$\cosh x = \dfrac{e^x + e^{-x}}{2}$, $\quad \sinh x = \dfrac{e^x - e^{-x}}{2}$, $\quad \tanh x = \dfrac{\sinh x}{\cosh x}$

$\cosh x = \dfrac{1}{\tanh x}$, $\quad \text{sech}\,x = \dfrac{1}{\cosh x}$, $\quad \text{cosech}\,x = \dfrac{1}{\sinh x}$

$\sin jx = j\sinh x$, $\cos jx = \cosh x$

$\sinh jx = j\sin x$, $\cosh jx = \cos x$

$\sin(x \pm jy) = \sin x\cosh y \pm j\cos x\sinh y$

$\cos(x \pm jy) = \cos x\cosh y \pm j\sin x\sinh y$

$\sinh(x \pm y) = \sinh x \cosh y \pm \cosh x \sinh y$

$\cosh(x \pm y) = \cos x \cosh y \pm j \sin x \sinh y$

$\sinh(x \pm jy) = \sinh x \cos y \pm j \cosh x \sin y$

$\cosh(x \pm jy) = \cosh x \cos y \pm j \sinh x \sin y$

$\tanh(x \pm jy) = \dfrac{\sinh 2x}{\cosh 2x + \cos 2y} \pm j \dfrac{\sin 2y}{\cosh 2x + \cos 2y}$

$\cosh^2 x - \sinh^2 x = 1$

$\mathrm{sech}^2 x + \tanh^2 x = 1$

C.5 Complex Variables

A complex number can be written as

$$z = x = jy = r\angle\theta = re^{j\theta} = r\,(\cos\theta + j\sin\theta),$$

where

$x = \mathrm{Re}\ z = r\cos\theta$

$y = \mathrm{Im}\ z = r\sin\theta$

$r = |z| = \sqrt{x^2 + y^2}, \quad \theta = \tan^{-1}\left(\dfrac{y}{x}\right)$

$j = \sqrt{-1}, \quad \dfrac{1}{j} = -j, \quad j^2 = -1$

The complex conjugate of $z = z^* = x - jy = r \angle -\theta = re^{-j\theta} = r(\cos\theta - j\sin\theta)$

$(e^{j\theta})^n = e^{jn\theta} = \cos n\theta + j\sin n\theta$ (de Moivre's theorem).

If $z_1 = x_1 + jy_1$ and $z_2 = x_2 + jy_2$, then only if $x_1 = x_2$ and $y_1 = y_2$,
$z_1 \pm z_2 = (x_1 + x_2) \pm j(y_1 + y_2)$

$z_1 z_2 = (x_1 x_2 - y_1 y_2) + j(x_1 y_2 + x_2 y_1) = r_1 r_2 e^{j(\theta_1 + \theta_2)} = r_1 r_2 \angle\theta_1 + \theta_2$

$\dfrac{z_1}{z_2} = \dfrac{(x_1 + jy_1)}{(x_2 + jy_2)} \cdot \dfrac{(x_2 + jy_2)}{(x_2 + jy_2)} = \dfrac{x_1 x_2 + y_1 y_2}{x_2^2 + y_2^2} + j\dfrac{x_2 y_1 - x_1 y_2}{x_2^2 + y_2^2}$

$= \dfrac{r_1}{r_2} e^{j(\theta_1 + \theta_2)} = \dfrac{r_1}{r_2} \angle\theta_1 - \theta_2$

$$\ln(re^{j\theta}) = \ln r + \ln e^{j\theta} = \ln r + j\theta + j2\ m\pi\ (m = \text{integer})$$

$$\sqrt{z} = \sqrt{x + jy} = \sqrt{r}\left(e^{j\theta/2}\right) = \sqrt{r}\angle\theta/2$$

$$z^n = (x + jy)^n = r^n\ e^{jn\theta} = r^n\angle n\theta\ (n = \text{integer})$$

$$z^{1/n} = (x + jy)^{1/n} = r^{1/n}\ e^{jn\theta} = r^{1/n}\angle\theta/2 + 2\ \pi m/n,\ (m = 0,1,2,\ldots,\ n-1)$$

C.6 Table of Derivatives

$y =$	$\dfrac{dy}{dx} =$
c (constant)	0
cx^n (n any constant)	cnx^{n-1}
e^{ax}	ae^{ax}
a^x ($a > 0$)	$a^x \ln a$
$\ln x$ ($x > 0$)	$\dfrac{1}{x}$
$\dfrac{c}{x^a}$	$\dfrac{-ca}{x^{a+1}}$
$\log_a x$	$\dfrac{\log_a e}{x}$
$\sin ax$	$a \cos ax$
$\cos ax$	$-a \sin ax$
$\tan ax$	$-a\sec^2 ax = \dfrac{a}{\cos^2 ax}$
$\cot ax$	$-a\operatorname{cosec}^2 ax = \dfrac{-a}{\sin^2 ax}$
$\sec ax$	$\dfrac{a\sin ax}{\cos^2 ax}$
$\operatorname{cosec} ax$	$\dfrac{-a\cos ax}{\sin^2 ax}$
$\arcsin ax = \sin^{-1} ax$	$\dfrac{a}{\sqrt{1 - a^2 x^2}}$
$\arccos ax = \cos^{-1} ax$	$\dfrac{-a}{\sqrt{1 - a^2 x^2}}$
$\arctan ax = \tan^{-1} ax$	$\dfrac{a}{1 + a^2 x^2}$
$\operatorname{arccot} ax = \cot^{-1} ax$	$\dfrac{-a}{1 + a^2 x^2}$
$\sinh ax$	$a \cosh ax$
$\cosh ax$	$a \sinh ax$

(Continued)

$y =$	$\dfrac{dy}{dx} =$
$\tanh ax$	$\dfrac{a}{\cosh^2 ax}$
$\sinh^{-1} ax$	$\dfrac{a}{\sqrt{1-a^2x^2}}$
$\cosh^{-1} ax$	$\dfrac{a}{\sqrt{a^2x^2-1}}$
$\tanh^{-1} ax$	$\dfrac{a}{1-a^2x^2}$
$u(x)+\upsilon(x)$	$\dfrac{du}{dx}+\dfrac{d\upsilon}{dx}$
$u(x)\,\upsilon(x)$	$u\dfrac{d\upsilon}{dx}+\upsilon\dfrac{du}{dx}$
$\dfrac{u(x)}{\upsilon(x)}$	$\dfrac{1}{\upsilon^2}\left(\upsilon\dfrac{du}{dx}-u\dfrac{d\upsilon}{dx}\right)$
$\dfrac{1}{\upsilon(x)}$	$\dfrac{-1}{\upsilon^2}\dfrac{d\upsilon}{dx}$
$y(\upsilon(x))$	$\dfrac{dy}{d\upsilon}\dfrac{d\upsilon}{dx}$
$y(\upsilon(u(x)))$	$=\dfrac{dy}{d\upsilon}\dfrac{d\upsilon}{du}\dfrac{du}{dx}$

C.7 Table of Integrals

$$\int a\,dx = ax+c \quad (c \text{ is an arbitrary constant})$$

$$\int x\,dy = xy - \int y\,dx$$

$$\int x^n dx = \frac{x^{n+1}}{n+1}+c, \quad (n \neq -1)$$

$$\int \frac{1}{x}dx = \ln|x|+c$$

$$\int e^{ax}dx = \frac{e^{ax}}{a}+c$$

$$\int a^x dx = \frac{a^x}{\ln a}+c \quad \text{for } (a > 0)$$

$$\int \ln x\, dx = x \ln x - x + c \quad \text{for } (x > 0)$$

$$\int \sin ax\, dx = \frac{-\cos ax}{a} + c$$

$$\int \cos ax\, dx = \frac{\sin ax}{a} + c$$

$$\int \tan ax\, dx = \frac{-\ln|\cos ax|}{a} + c$$

$$\int \cot ax\, dx = \frac{-\ln|\cos ax|}{a} + c$$

$$\int \sec ax\, dx = \frac{-\ln\left(\dfrac{1 - \sin ax}{1 + \sin ax}\right)}{2a} + c$$

$$\int \operatorname{cosec} ax\, dx = \frac{-\ln\left(\dfrac{1 - \cos ax}{1 + \cos ax}\right)}{2a} + c$$

$$\int \frac{1}{x^2 + a^2}\, dx = \frac{\tan^{-1}(x/a)}{a} + c$$

$$\int \frac{1}{x^2 - a^2}\, dx = \frac{\ln\big((x-a)/(x+a)\big)}{2a} + c \quad \text{or} \quad \frac{\tanh^{-1}(x/a)}{a} + c$$

$$\int \frac{1}{a^2 - x^2}\, dx = \frac{\ln\big((x+a)/(x-a)\big)}{2a} + c$$

$$\int \frac{1}{\sqrt{a^2 - x^2}}\, dx = \sin^{-1}\left(\frac{x}{a}\right) + c$$

$$\int \frac{1}{\sqrt{a^2 - x^2}}\, dx = \frac{\sinh^{-1}(x/a)}{a} + c \quad \text{or} \quad \ln\left(x + \sqrt{x^2 + a^2}\right) + c$$

$$\int \frac{1}{\sqrt{x^2 - a^2}}\, dx = \ln\left(x + \sqrt{x^2 + a^2}\right) + c$$

$$\int \frac{1}{x\sqrt{x^2 - a^2}}\, dx = \frac{\sec^{-1}(x/a)}{a} + c$$

$$\int xe^{ax}\, dx = \frac{(ax - 1)e^{ax}}{a^2} + c$$

$$\int x \cos ax\, dx = \frac{\cos ax + ax \sin ax}{a^2} + c$$

$$\int x \sin axdx = \frac{\sin ax + ax \cos ax}{a^2} + c$$

$$\int x \ln xdx = \frac{x^2}{2} \ln x - \frac{x^2}{4} + c$$

$$\int xe^{ax} dx = \frac{e^{ax}(ax-1)}{a^2} + c$$

$$\int e^{ax} \cos bxdx = \frac{e^{ax}(a\cos bx + b\sin bx)}{a^2 + b^2} + c$$

$$\int e^{ax} \sin bxdx = \frac{e^{ax}(-b\cos bx + a\sin bx)}{a^2 + b^2} + c$$

$$\int \sin^2 xdx = \frac{x}{2} - \frac{\sin 2x}{4} + c$$

$$\int \cos^2 xdx = \frac{x}{2} - \frac{\sin 2x}{4} + c$$

$$\int \tan^2 xdx = \tan x - x + c$$

$$\int \cot^2 xdx = \cot x - x + c$$

$$\int \sec^2 xdx = \tan x + c$$

$$\int \mathrm{cosec}^2 xdx = -\cot x + c$$

$$\int \sec x \tan xdx = -\sec x + c$$

$$\int \mathrm{cosec}\, x \cot xdx = -\mathrm{cosec}\, x + c$$

C.8 Table of Probability Distributions

1. Discrete Distribution	Probability $P(X=x)$	Expectation (Mean) μ	Variance σ^2
Binomial $B(n, p)$	$\binom{n}{r} p^r (1-p)^{n-r} = \frac{n! p^r q^{n-1}}{r!(n-r)!} np$ $r = 0, 1, \ldots, n$	Np	$np(1-p)$
Geometric $G(p)$	$(1-p)^{r-1} p$	$\dfrac{1}{p}$	$\dfrac{1-p}{p^2}$

(*Continued*)

1. Discrete Distribution	Probability $P(X=x)$	Expectation (Mean) μ	Variance σ^2
Poisson $p(\lambda)$	$\dfrac{\lambda^n e^{-\lambda}}{n!}$	λ	λ
Pascal (negative binomial) $NB(r, p)$	$\begin{pmatrix} x & -1 \\ r & -1 \end{pmatrix} p^r (1-p)^{x-r},$ $x = r, r+1, \ldots$	$\dfrac{r}{p}$	$\dfrac{r(1-p)}{p^2}$
Hypergeometric $H(N, n, p)$	$\dfrac{\begin{pmatrix} Np \\ r \end{pmatrix}\begin{pmatrix} N-Np \\ n-r \end{pmatrix}}{\begin{pmatrix} N \\ n \end{pmatrix}}$	Np	$np(1-p)\dfrac{N-n}{N-1}$

2. Discrete Distribution	Density $f(x)$	Expectation (Mean) μ	Variance σ^2
Exponential $E(\lambda)$	$\begin{cases} \lambda e^{-\lambda x}, & x \geq 0 \\ 0 & x < 0 \end{cases}$	$\dfrac{1}{\lambda}$	$\dfrac{1}{\lambda^2}$
Uniform $U(a, b)$	$\begin{cases} \dfrac{1}{b-a}, & a < x < b \\ 0, & \text{elsewhere} \end{cases}$	$\dfrac{a+b}{2}$	$\dfrac{(b-a)^2}{12}$
Standardized normal $N(0, 1)$	$\varphi(x) = \dfrac{e^{-x^2/2}}{\sqrt{2\pi}}$	0	1
General normal	$\dfrac{1}{\sigma}\varphi\left(\dfrac{x-\mu}{\sigma}\right)$	μ	σ^2
Gamma $\Gamma(n,\lambda)$	$\dfrac{\lambda^n}{\Gamma(n)}x^{n-1}e^{-\lambda x}$	$\dfrac{n}{\lambda}$	$\dfrac{n}{\lambda^2}$
Beta $\beta(p, q)$	$a_{p,q}x^{p-1}(1-x)^{q-1}, \quad 0 \leq x \leq 1$ $a_{p,q} = \dfrac{\Gamma(p+q)}{\Gamma(p)\Gamma(q)}, \quad p > 0, q > 0$	$\dfrac{p}{p+q}$	$\dfrac{pq}{(p+q)^2(p+q+1)}$
Weibull $W(\lambda, \beta)$	$\lambda^\beta \beta x^{\beta-1}e^{-(\lambda x)\beta}, \quad x \geq 0$ $F(x) = 1 - e^{-(\lambda x)\beta}$	$\dfrac{1}{\lambda}\Gamma\left(1+\dfrac{1}{\beta}\right)$	$\dfrac{1}{\lambda^2}(A-B)$ $A = \Gamma^2\left(1+\dfrac{2}{\beta}\right)$ $B = \Gamma^2\left(1+\dfrac{1}{\beta}\right)$
Rayleigh $R(\sigma)$	$\dfrac{x}{\sigma^2}e^{-x^2/2\sigma^2}, \quad x \geq 0$	$\sigma\sqrt{\dfrac{\pi}{2}}$	$2\sigma^2\left(1-\dfrac{\pi}{4}\right)$

C.9 Summations (Series)

C.9.1 Finite Element of Terms

$$\sum_{n=0}^{N} a^n = \frac{1-a^{N+1}}{1-a}; \quad \sum_{n=0}^{N} na^n = a\left(\frac{1-(N+1)a^N + Na^{N+1}}{(1-a)^2}\right)$$

$$\sum_{n=0}^{N} n = \frac{N(N+1)}{2}; \quad \sum_{n=0}^{N} n^2 = \frac{N(N+1)(2N+1)}{6}$$

$$\sum_{n=0}^{N} n(n+1) = \frac{N(N+1)(N+2)}{3};$$

$$(a+b)^N = \sum_{n=0}^{N} NC_n a^{N-n} b^n, \quad \text{where } NC_n = NC_{N-n} = \frac{NP_n}{n!} = \frac{N!}{(N-n)!n!}$$

C.9.2 Infinite Element of Terms

$$\sum_{n=0}^{\infty} x^n = \frac{1}{1-x}, (|x|<1); \quad \sum_{n=0}^{\infty} nx^n = \frac{1}{(1-x)^2}, (|x|<1)$$

$$\sum_{n=0}^{\infty} n^k x^n = \lim_{a\to 0}(-1)^k \frac{\partial^k}{\partial a^k}\left(\frac{x}{x-e^{-a}}\right), (|x|<1); \quad \sum_{n=0}^{\infty} \frac{(-1)^n}{2n+1} = 1 - \frac{1}{3} + \frac{1}{5} - \frac{1}{7} + \cdots = \frac{1}{4}\pi$$

$$\sum_{n=0}^{\infty} \frac{1}{n^2} = 1 + \frac{1}{2^2} + \frac{1}{3^2} + \frac{1}{4^2} + \cdots = \frac{1}{6}\pi^2$$

$$e^x = \sum_{n=0}^{\infty} \frac{x^n}{n!} = 1 + \frac{1}{1!}x + \frac{1}{2!}x^2 + \frac{1}{3!}x^3 + \cdots$$

$$a^x = \sum_{n=0}^{\infty} \frac{(\ln a)^n x^n}{n!} = 1 + \frac{(\ln a)x}{1!} + \frac{(\ln a)^2 x^2}{2!} + \frac{(\ln a)^3 x^3}{3!} + \cdots$$

$$\ln(1\pm x) = \sum_{n=0}^{\infty} \frac{(\pm 1)^n x^x}{n} = \pm x - \frac{x^2}{2} \pm \frac{x^3}{3} - \cdots, \quad (|x|<1)$$

$$\sin x = \sum_{n=0}^{\infty} \frac{(-1)^n x^{2n+1}}{(2n+1)!} = x - \frac{x^3}{3!} + \frac{x^5}{5!} - \frac{x^7}{7!} + \cdots$$

$$\cos x = \sum_{n=0}^{\infty} \frac{(-1)^n x^{2n}}{(2n)!} = 1 - \frac{x^2}{2!} + \frac{x^4}{4!} - \frac{x^6}{6!} + \cdots$$

$$\tan x = x + \frac{x^3}{3} + \frac{2x^5}{15} + \cdots, \quad (|x| < 1)$$

$$\tan^{-1} x = \sum_{n=0}^{\infty} \frac{(-1)^n x^{2n+1}}{(2n+1)} = x - \frac{x^2}{3} + \frac{x^5}{5} - \frac{x^7}{7} + \cdots, \quad (|x| < 1)$$

C.10 Logarithmic Identities

$\log_e a = \ln a$ (natural logarithm)

$\log_{10} a = \log a$ (common logarithm)

$\log ab = \log a + \log b$

$\log \dfrac{a}{b} = \log a - \log b$

$\log a^n = n \log a$

C.11 Exponential Identities

$$e^x = 1 + x + \frac{x^2}{2!} + \frac{x^3}{3!} + \frac{x^4}{4!} + \cdots, \quad \text{where } e \simeq 2.7182$$

$e^x e^y = e^{x+y}$

$(e^x)^n = e^{nx}$

$\ln e^x = x$

C.12 Approximations for Small Quantities

If $|a| \ll 1$, then

$\ln (1+a) = a$

$e^a = 1 + a$

$\sin a = a$

$\cos a = 1$

$\tan a = a$

$(1 \pm a)^n = 1 \pm na$

C.13 Matrix Notation and Operations

C.13.1 Matrices

A *matrix* is a rectangular array of elements arranged in rows and columns. The array is commonly enclosed in brackets. Let a matrix A (expressed in boldface as $\mathbf{A}$ or in bracket as [A]) have m rows and n columns; then the matrix can be expressed by

$$\mathbf{A} = [A] = \begin{bmatrix} a_{11} & a_{12} & . & . & . & a_{1j} & . & . & . & a_{1n} \\ a_{21} & a_{22} & . & . & . & a_{2j} & . & . & . & a_{2n} \\ . & . & . & . & . & . & . & . & . \\ . & . & . & . & . & . & . & . & . \\ . & . & . & . & . & . & . & . & . \\ a_{i1} & a_{i2} & . & . & . & a_{ij} & . & . & . & a_{in} \\ . & . & . & . & . & . & . & . & . \\ . & . & . & . & . & . & . & . & . \\ . & . & . & . & . & . & . & . & . \\ a_{m1} & a_{m2} & . & . & . & a_{mj} & . & . & . & a_{mn} \end{bmatrix}$$

where the element a_{ij} has two subscripts, the first referring to the row position of the element in the array and the second the column position. A matrix with m rows and n columns, [A], is defined as a matrix of order or size $m \times n$ (m by n), or an $m \times n$ matrix. A vector is a matrix that consists of only one row or one column. *Location of an element in a matrix:*

$$\text{Let } A = \begin{bmatrix} a_{11} & a_{12} & a_{13} & a_{14} \\ a_{21} & a_{22} & a_{23} & a_{24} \\ a_{31} & a_{32} & a_{33} & a_{34} \\ a_{41} & a_{42} & a_{43} & a_{44} \end{bmatrix} \text{ is a matrix with size } 4 \times 4$$

where
a_{11} is the element a at row 1 and column 1
a_{12} is the element a at row 1 and column 2
a_{32} is the element a at row 3 and column 2

C.13.2 Special Common Types of Matrices

1. If $m \pm n$, then the matrix $[A]$ is called rectangular matrix.
2. If $m = n$, then the matrix $[A]$ is called square matrix of order n.
3. If $m = 1$ *and* $n > 1$, then the matrix $[A]$ is called row matrix or row vector.
4. If $m > 1$ *and* $n = 1$, then the matrix $[A]$ is called column matrix or column vector.
5. If $m = 1$ *and* $n = 1$, then the matrix $[A]$ is called a scalar.
6. A *real matrix* is a matrix whose elements are all real.
7. A *complex matrix* is a matrix whose elements may be complex.
8. A *null matrix* is a matrix whose elements are all zero.
9. An *identity* (or *unit*) *matrix*, $[I]$ or **I**, *is a square matrix whose elements are equal to zero except those located on its main diagonal* elements, which are unity (or one). *Main diagonal* elements have equal row and column subscripts. The main diagonal runs from the upper-left corner to the lower-right corner. If the elements of an identity matrix are denoted as e_{ij}, then

$$e_{ij} = \begin{cases} 1 & i = j \\ 0 & i \neq j \end{cases}$$

10. A *diagonal matrix* is a square matrix that has zero elements everywhere except on its main diagonal. That is, for diagonal matrix, $a_{ij} = 0$ when $i \pm j$ and not all a_{ii} are zero.
11. A *symmetric matrix* is a square matrix whose elements satisfy the condition $a_{ij} = -a_{ji}$ for $i \neq j$, and $a_{ii} = 0$.
12. An *antisymmetric* (or *skew symmetric*) *matrix* is a square matrix whose elements $a_{ij} = -a_{ji}$ for $i \neq j$, and $a_{ii} = 0$.
13. A *triangular matrix* is a square matrix whose all elements on one side of the diagonal are zero. There are two types of triangular matrices: first, an upper triangular **U**, whose elements below the diagonal are zero, and second, a lower triangular **L**, whose elements above the diagonal are all zero.
14. A *partitioned* (or *block*) *matrix* is a matrix that is divided by horizontal and vertical lines into smaller matrices called submatrics or blocks.

C.13.3 Matrix Operations

1. Transpose of a matrix

   The *transpose* of a matrix $\mathbf{A} = [a]$ is denoted as $\mathbf{AT} = [a, J]$ and is obtained by interchanging the rows and columns in matrix $\mathbf{A}$. Thus, if a matrix $\mathbf{A}$ is of order $m \times n$, then $\mathbf{AT}$ will be of order $n \times m$.

2. Addition and subtraction

   Addition and subtraction can only be performed for matrices of the same size. The addition is accomplished by adding corresponding elements of each matrix. For addition, $\mathbf{C} = \mathbf{A} + \mathbf{B}$ implies that $c_{ij} = a_{ij} + b_{ij}$.

   Now, the subtraction is accomplished by subtracting corresponding elements of each matrix. For subtraction, $\mathbf{C} = \mathbf{A} - \mathbf{B}$ implies that $c_{ij} = a_{ij} - b_{ij}$, where c_{ij}, a_{ij}, and b_{ij}, are typical elements of the $\mathbf{C}$, $\mathbf{A}$, and $\mathbf{B}$ matrices, respectively.

   If matrices $\mathbf{A}$ and $\mathbf{B}$ are both of the same size $m \times n$, the resulting matrix $\mathbf{C}$ is also of size $m \times n$.

   Matrix addition and subtraction are associative:

   $$\mathbf{A} + \mathbf{B} + \mathbf{C} = (\mathbf{A} + \mathbf{B}) + \mathbf{C} = \mathbf{A} + (\mathbf{B} + \mathbf{C})$$

   $$\mathbf{A} + \mathbf{B} - \mathbf{C} = (\mathbf{A} + \mathbf{B}) - \mathbf{C} = \mathbf{A} + (\mathbf{B} - \mathbf{C})$$

   Matrix addition and subtraction are commutative:

   $$\mathbf{A} + \mathbf{B} = \mathbf{B} + \mathbf{A}$$

   $$\mathbf{A} - \mathbf{B} = -\mathbf{B} + \mathbf{A}$$

3. Multiplication by scalar

   A matrix is multiplied by a scalar by multiplying each element of the matrix by the scalar. The multiplication of a matrix $\mathbf{A}$ by a scalar c is defined as

   $$c\mathbf{A} = [ca_{ij}]$$

   The scalar multiplication is commutative.

4. Matrix multiplication

   The product of two matrices is $\mathbf{C} = \mathbf{AB}$ if and only if the number of columns in $\mathbf{A}$ is equal to the number of rows in $\mathbf{B}$. The product of matrix $\mathbf{A}$ of size $m \times n$ and matrix $\mathbf{B}$ of size $n \times r$ results in matrix $\mathbf{C}$ of size $m \times r$. Then, $c_{ij} = \sum_{k=1}^{n} a_{ik} b_{kj}$

That is, the (ij)th component of matrix $\mathbf{C}$ is obtained by taking the dot product

$$c_{ij} = (i\text{th row of } \mathbf{A}) \cdot (j\text{th column of } \mathbf{B})$$

Matrix multiplication is associative:

$$\mathbf{ABC} = (\mathbf{AB})\mathbf{C} = \mathbf{A}(\mathbf{BC})$$

Matrix multiplication is distributive:

$$\mathbf{A}\,(\mathbf{B} + \mathbf{C}) = \mathbf{AB} + \mathbf{AC}$$

Matrix multiplication is not commutative:

$$\mathbf{AB} \neq \mathbf{BA}$$

5. Transpose of matrix multiplication

The transpose of matrix multiplication is usually denoted $(\mathbf{AB})^T$ and is defined as

$$(\mathbf{AB})^T = \mathbf{B}^T \mathbf{A}^T$$

6. Inverse of square matrix

The inverse of a matrix $\mathbf{A}$ is denoted by $\mathbf{A}^{-1}$. The inverse matrix satisfies

$$\mathbf{A}\mathbf{A}^{-1} = \mathbf{A}^{-1}\mathbf{A} = \mathbf{I}$$

A matrix that possesses an inverse is called nonsingular matrix (or invertible matrix). A matrix without an inverse is called a singular matrix.

7. Differentiation of a matrix

The differentiation of a matrix is differentiation of every element of the matrix separately. To emphasize, if the elements of the matrix $\mathbf{A}$ are a function of t, then

$$\frac{d\mathbf{A}}{dt} = \left[\frac{da_{ij}}{dt} \right]$$

8. Integration of a matrix

The integration of a matrix is integration of every element of the matrix separately. To emphasize, if the elements of the matrix **A** are a function of t, then

$$\int \mathbf{A} dt = \left[\int a_{ij} dt \right]$$

9. Equality of matrices

Two matrices are equal if they have the same size and their corresponding elements are equal.

C.13.4 Determinant of a Matrix

The determinant of a square matrix **A** is a scalar number denoted by $|\mathbf{A}|$ or det **A**.

The value of a second-order determinant is calculated from

$$\det \begin{bmatrix} a_{11} & a_{12} \\ a_{21} & a_{22} \end{bmatrix} = \begin{vmatrix} a_{11} & a_{12} \\ a_{21} & a_{22} \end{vmatrix} = a_{11}a_{22} - a_{12}a_{21}$$

By using the sign rule of each term, the determinant is determined by the first row in the diagram:

$$\begin{vmatrix} + & - & + \\ - & + & - \\ + & - & + \end{vmatrix}$$

The value of a third-order determinant is calculated as

$$\det \begin{bmatrix} a_{11} & a_{12} & a_{13} \\ a_{21} & a_{22} & a_{23} \\ a_{31} & a_{32} & a_{33} \end{bmatrix} = \begin{bmatrix} a_{11} & a_{12} & a_{13} \\ a_{21} & a_{22} & a_{23} \\ a_{31} & a_{32} & a_{33} \end{bmatrix}$$

$$= a_{11} \begin{vmatrix} a_{22} & a_{23} \\ a_{32} & a_{33} \end{vmatrix} - a_{12} \begin{vmatrix} a_{21} & a_{23} \\ a_{31} & a_{33} \end{vmatrix} + a_{13} \begin{vmatrix} a_{21} & a_{22} \\ a_{31} & a_{32} \end{vmatrix}$$

C.14 Vectors

C.14.1 Vector Derivative

1. Cartesian coordinates

Coordinates	(x, y, z)
Vector	$A = A_x a_x + A_y a_y + A_z a_z$
Gradient	$\nabla A = \dfrac{\partial A}{\partial x} a_x + \dfrac{\partial A}{\partial y} a_y + \dfrac{\partial A}{\partial z} a_z$
Divergence	$\nabla \cdot A = \dfrac{\partial A_x}{\partial x} + \dfrac{\partial A_y}{\partial y} + \dfrac{\partial A_z}{\partial z}$
Curl	$\nabla \times A = \begin{vmatrix} a_x & a_y & a_z \\ \dfrac{\partial}{\partial x} & \dfrac{\partial}{\partial y} & \dfrac{\partial}{\partial z} \\ A_x & A_y & A_z \end{vmatrix}$
	$= \left(\dfrac{\partial A_z}{\partial y} - \dfrac{\partial A_y}{\partial z} \right) a_x + \left(\dfrac{\partial A_x}{\partial z} - \dfrac{\partial A_z}{\partial x} \right) a_y + \left(\dfrac{\partial A_y}{\partial x} - \dfrac{\partial A_x}{\partial y} \right) a_z$
Laplacian	$\nabla^2 A = \dfrac{\partial^2 A}{\partial x^2} + \dfrac{\partial^2 A}{\partial y^2} + \dfrac{\partial^2 A}{\partial z^2}$

2. Cylindrical coordinates

Coordinates	(ρ, ϕ, z)
Vector	$A = A_\rho a_\rho + A_\phi a_\phi + A_z a_z$
Gradient	$\nabla A = \dfrac{\partial A}{\partial \rho} a_\rho + \dfrac{1}{\rho} \dfrac{\partial A}{\partial \phi} a_\phi + \dfrac{\partial A}{\partial z} a_z$
Divergence	$\nabla \cdot A = \dfrac{1}{\rho} \dfrac{\partial}{\partial \rho} (\rho A_\rho) + \dfrac{\partial A_\phi}{\partial \phi} + \dfrac{\partial A_z}{\partial z}$
Curl	$\nabla \times A = \begin{vmatrix} a_\rho & \rho a_\phi & a_z \\ \dfrac{\partial}{\partial \rho} & \dfrac{\partial}{\partial \phi} & \dfrac{\partial}{\partial z} \\ A_\rho & \rho A_\phi & A_z \end{vmatrix}$
	$= \left(\dfrac{1}{\rho} \dfrac{\partial A_z}{\partial \phi} - \dfrac{\partial A_\phi}{\partial z} \right) a_\rho + \left(\dfrac{\partial A_\rho}{\partial z} - \dfrac{\partial A_z}{\partial \rho} \right) a_\phi + \dfrac{1}{\rho} \left(\dfrac{\partial}{\partial x} (\rho A_\phi) - \dfrac{\partial A_\rho}{\partial \rho} \right) a_z$
Laplacian	$\nabla^2 A = \dfrac{1}{\rho} \dfrac{\partial}{\partial \rho} \left(\rho \dfrac{\partial A}{\partial \rho} \right) + \dfrac{1}{\rho^2} \dfrac{\partial^2 A}{\partial \phi^2} + \dfrac{\partial^2 A}{\partial z^2}$

3. Spherical coordinates

Coordinates	(r, θ, ϕ)
Vector	$A = A_r a_r + A_\theta a_\theta + A_\phi a_\phi$
Gradient	$\nabla A = \dfrac{\partial A}{\partial r} a_r + \dfrac{1}{r}\dfrac{\partial A}{\partial \theta} a_\theta + \dfrac{1}{r\sin\theta}\dfrac{\partial A}{\partial \phi} a_\phi$
Divergence	$\nabla \cdot A = \dfrac{1}{r^2}\dfrac{\partial}{\partial r}\left(r^2 A_r\right) + \dfrac{1}{r\sin\theta}\dfrac{\partial}{\partial \theta}\left(A_\theta \sin\theta\right) + \dfrac{1}{r\sin\theta}\dfrac{\partial A_\phi}{\partial \phi}$

Curl

$$\nabla \times A = \frac{1}{r^2\sin\theta}\begin{vmatrix} a_r & r a_\theta & (r\sin\theta)a_\phi \\ \dfrac{\partial}{\partial r} & \dfrac{\partial}{\partial \theta} & \dfrac{\partial}{\partial \phi} \\ A_r & r A_\theta & (r\sin\theta)A_\phi \end{vmatrix}$$

$$= \frac{1}{r\sin\theta}\left(\frac{\partial}{\partial \theta}\left(A_\phi \sin\theta\right) - \frac{\partial A_{\theta\phi}}{\partial \phi}\right)a_r + \frac{1}{r}\left(\frac{1}{\sin\theta}\frac{\partial A_r}{\partial \phi} - \frac{\partial}{\partial r}\left(r A_\phi\right)\right)a_\theta$$

$$+ \frac{1}{r}\left(\frac{\partial}{\partial r}\left(r A_\theta\right) - \frac{\partial A_r}{\partial \theta}\right)a_\phi$$

Laplacian

$$\nabla^2 A = \frac{1}{r^2}\frac{\partial}{\partial r}\left(r^2 \frac{\partial A}{\partial r}\right) + \frac{1}{r^2\sin\theta}\frac{\partial}{\partial \theta}\left(\sin\theta \frac{\partial A}{\partial \theta}\right) + \frac{1}{r^2\sin\theta}\frac{\partial^2 A}{\partial \phi^2}$$

C.14.2 Vector Identity

1. Triple products

$$\mathbf{A}\cdot(\mathbf{B}\times\mathbf{C}) = \mathbf{B}\cdot(\mathbf{C}\times\mathbf{A}) = \mathbf{C}\cdot(\mathbf{A}\times\mathbf{B})$$

$$\mathbf{A}\cdot(\mathbf{B}\times\mathbf{C}) = \mathbf{B}(\mathbf{A}\cdot\mathbf{C}) - \mathbf{C}(\mathbf{A}\cdot\mathbf{B})$$

2. Product rules

$$\nabla(fg) = f(\nabla g) + g(\nabla f)$$

$$\nabla(\mathbf{A}\cdot\mathbf{B}) = \mathbf{A}\times(\nabla\times\mathbf{B}) + \mathbf{B}\times(\nabla\times\mathbf{A}) + (\mathbf{A}\cdot\nabla)\mathbf{B} + (\mathbf{B}\times\nabla)\mathbf{A}$$

$$\nabla\cdot(f\mathbf{A}) = f(\nabla\cdot\mathbf{A}) + \mathbf{A}\cdot(\nabla f)$$

$$\nabla(\mathbf{A}\times\mathbf{B}) = \mathbf{B}\cdot(\nabla\times\mathbf{A}) - \mathbf{A}\cdot(\nabla\times\mathbf{B})$$

$$\nabla\times(f\mathbf{A}) = f(\nabla\times\mathbf{A}) - \mathbf{A}\times(\nabla f) = \nabla\times(f\mathbf{A}) = f(\nabla\times\mathbf{A}) + (\nabla f)\times\mathbf{A}$$

$$\nabla\times(\mathbf{A}\times\mathbf{B}) = (\mathbf{B}\cdot\nabla)\mathbf{A} - (\mathbf{A}\cdot\nabla)\mathbf{B} + \mathbf{A}(\nabla\cdot\mathbf{B}) - (\nabla\cdot\mathbf{A})$$

3. Second derivative

$$\nabla \cdot (\nabla \times \mathbf{A}) = 0$$

$$\nabla \times (\nabla f) = 0$$

$$\nabla \cdot (\nabla f) = \nabla^2 f$$

$$\nabla \times (\nabla \times A) = \nabla(\nabla \cdot \mathbf{A}) - \nabla^2 \mathbf{A}$$

4. Addition, division, and power rules

$$\nabla(f + g) = \nabla f + \nabla g$$

$$\nabla \cdot (\mathbf{A} + \mathbf{B}) = \nabla \cdot \mathbf{A} + \nabla \cdot \mathbf{B}$$

$$\nabla \times (\mathbf{A} \times \mathbf{B}) = \nabla \times \mathbf{A} + \nabla \times \mathbf{B}$$

$$\nabla \left(\frac{f}{g} \right) = \frac{g(\nabla f) - f(\nabla g)}{g^2}$$

$$\nabla f^n = n f^{n-1} \nabla f \quad (n = \text{integer})$$

C.14.3 Fundamental Theorems

1. Gradient theorem

$$\int_a^b (\nabla f) \cdot dl = f(b) - f(a)$$

2. Divergence theorem

$$\int_{\text{volume}} (\nabla \cdot \mathbf{A}) \cdot dv = \oint_{\text{surface}} \mathbf{A} \cdot ds$$

3. Curl (Stokes) theorem

$$\int_{\text{surface}} (\nabla \times \mathbf{A}) \cdot ds = \oint_{\text{line}} \mathbf{A} \cdot dl$$

4. $$\oint_{\text{line}} f dl = - \oint_{\text{surface}} \nabla f \times ds$$

5. $$\oint_{\text{surface}} f dls = - \oint_{\text{volume}} \nabla f dv$$

6. $$\oint_{\text{surface}} \mathbf{A} \times ds = - \int_{\text{volume}} \nabla \times \mathbf{A} dv$$

C.15 Periodic Table

Index

A

Al$_2$O$_3$–water nanofluids
  anomalous thermal conductivity
    enhancement, 102
  Ansys Fluent
    computational domain, 112
    grid independent test,
      112–113
    Poiseuille numbers, 112–113
  boiling heat transfer, 11–14
  boundary conditions, 110–111
  Brownian motion, 104
  Brownian velocity, 109
  convective heat transfer, 102–104
  effective mixture density, 109
  elliptic cross section tubes, 105,
    114–115
  friction factor, 114
  Mirmasoumi and
    Behzadmehr cases, 111
  mixture model
    drift velocity, 107
    geometry and coordinate
      system, 106
    relative velocity, 108
    slip velocity, 107
    steady-state equations, 107
  mixture theory, 105
  multiphase flow
    definition, 102
    flow map, 102–103
    flow regimes, 102–103
  nanofluid viscosity, 109
  Nusselt number, 114–115
  physical quantities, 111
  research opportunities, 115–116
  thermal conductivity, 110
  thermal expansion coefficient, 110
  thermophoresis, 104
  thermophysical properties, 108
  viscous properties, 7

Aluminum nitride (AlN)
  nanopowder, 88
Atmospheric pressure ionization
  mass spectrometry (API-MS),
  145–146

B

Brownian motion, 104
Brownian velocity, 109

C

Cancer biomarker, 145
Cancer management/detection
  API-MS, 145–146
  biomarker, 145
  drug effects, 126
  high-resolution analyzers, 146
  imaging, impact of, 125
  LC-MS/MS, 146
  mass spectrometers and techniques,
    146–147
  mechanisms, imaging modalities,
    142–143
  molecular imaging, 141–142
  MRI, 142, 144
  nanoparticles, 140
  optical imaging, 142–143
  pharmaceutical formulations, 127
  proteomics cancer research, 146, 148
  QDs
    applications, 124
    breast cancer, 148
    carbohydrate antigen 125, 147–148
    lymph nodes metastasis, 149
    optical and electrical
      properties, 140
    optical imaging, 143–144
    silk fibroin, 148
  smart nanoparticle, 125
  SPIONs, 144

theranostics, 125
tumor angiogenesis, 125–127
tumor biomarkers, 146
Carbohydrate antigen 125 (CA 125),
    147–148
Carbon nanotubes (CNTs)
    nucleate pool boiling, 12–13
    surface tension, 8
    thermal conductivity, 5
    viscous properties, 7–8
CHFs, *see* Critical heat fluxes (CHFs)
CNTs, *see* Carbon nanotubes (CNTs)
Complex variables, 218–219
Computational flow analysis
    advantages, 182–183
    biomedical data, 185
    clarification and prediction, 189
    clinical microscopy, 184
    diagnosis and therapy, 190
    Doppler ultrasonography, 191–192
    flow cytometry, 192–194
    MATLAB® program, 190–191
    nanoneedles, 194–195
    in silico approach, 174
Computational fluid dynamics (CFD), 104
Convective heat transfer, nanofluids
    fractal theory, 44–46
    Monte Carlo simulations
        algorithm, 47
        convergence criterion, 47
        cumulative probability, 46
        vs. experimental data, 48–49
        heat flux *vs.* average diameter of
           nanoparticles, 50
        mean nanoparticle diameter, 46
        variance, 48
        volumetric nanoparticle
           concentration, 48
Copper(II) oxide (CuO)
    nanopowder, 88
Critical heat fluxes (CHFs)
    bubble departure diameter, 53
    diameter of nanoparticles, 52
    flow boiling heat transfer
        CuO–water nanofluids, 14–15
        fabrication technology, 16
        particle size's effect, 15
        vs. pool boiling process, 15, 21–22

fractal dimension, 54–55
    minimum and maximum active
        cavity diameters, 53
    nucleate pool boiling heat transfer
        Al_2O_3–water nanofluids, 11–14
        γ-Al_2O_3–water nanofluids, 12
        CNTs, 12–13
        CuO/water nanofluids, 11
        vs. flow boiling heat transfer, 15,
           21–22
        refrigerant R-141b, 10–11
        SiO_2–water nanofluids, 14
        ZrO_2–water nanofluids, 12
    pool boiling heat transfer *vs.*
        average diameter of
        nanoparticles, 58
    predicted fractal models *vs.* existing
        experimental data, Al_2O_3
        nanofluids, 57
    present model predictions *vs.*
        experimental data
        Al_2O_3 nanofluids, 55–56
        SiO_2 nanofluids, 56–57
        TiO_2 nanofluids, 55–56
    SEM micrographs, 51
    volumetric nanoparticle
        concentration, 59
    wall superheat, 55
    water-based nanoparticle
        suspensions, 56
Crosser model, 96–97
CuO–water nanofluids
    flow boiling heat transfer, 14–15
    nucleate pool boiling heat
        transfer, 11
    viscous properties, 6–7
Curl (Stokes) theorem, 233

D

Degree of linear polarization
    (DOLP), 150
Degree of polarization (DOP), 150
Dendrimers, 139
Derivatives table, 219–220
Divergence theorem, 233
DOLP, *see* Degree of linear polarization
    (DOLP)

DOP, *see* Degree of polarization (DOP)
Doppler ultrasonography, 191–192

E

Electromagnetic spectrum, 210–211
Enhanced permeability and retention
(EPR), 135
Ethylene glycol (EG), nanofluid
preparation, 88–90
Exponential identities, 225

F

Figure of merits (FOMs), 152
Finite element summations, 224
Flow analysis
body fluids, 175
computational technology
advantages, 182–183
biomedical data, 185
clarification and prediction, 189
clinical microscopy, 184
diagnosis and therapy, 190
Doppler ultrasonography, 191–192
flow cytometry, 192–194
MATLAB® program, 190–191
nanoneedles, 194–195
in silico approach, 174
fluid dynamics
compressibility, 176–178
conservation laws, 176–177
laminar and turbulent, 176, 179–180
magnetohydrodynamics, 177,
180–181
steady-state flow, 176, 178–179
viscous properties, 176, 178
micro and nanoflows
advantages, 185
computational tools, 188–189
microfluidics, 186–188
nanofluidics, 187–188
nanoscience technology, 173
problems and disorders, 185–186
physical properties, 172
states of matter, 172
Flow boiling heat transfer
CuO–water nanofluids, 14–15

fabrication technology, 16
industrial applications, 2
in microchannels, 2
particle size's effect, 15
vs. pool boiling process, 15, 21–22
Fluid dynamics
assumptions, 129
bistable switch, 133–134
drug transport, 132–133
equations, 129–130
information flow, 129–130
monostable switch, 134
symbols definition, 129, 131
tumor blood flow, 131–132
tumor cell density dynamics, 134
tumor microenvironment, 128–129
tumor vessel, 128
Fluid mechanics, 122
FOMs, *see* Figure of merits (FOMs)
Fractal dimension, 33

G

Gold nanoparticles, 124
Gradient theorem, 233
Greek alphabet, 215

H

Hamilton and Crosser model, 96–97
Hot-wire calorimeter, 80–81
Hyperbolic functions, 217–218

I

Index of refraction, 209
Infinite element summations, 224–225
Integrals table, 220–222

J

Joule's effect, 81

L

Lipid nanoparticles, 137–138
Liquid chromatography-tandem mass
spectrometry (LC-MS/MS), 146
Logarithmic identities, 225

M

Magnetohydrodynamics, 177, 180–181
Material constants
    conductivity, 205, 207
    dielectric constants, 206–207
    dielectric strength, 206
    relative permeability, 206
MATLAB® program, 190–191
Matrices, 226–230
Maxwell, Hamilton, and Crosser
        model, 96
Maxwell model, 96
Mean squared error (MSE), 152
Metabolomics, 123
Microfluidics, 186–188
Mixture model
    drift velocity, 107
    geometry and coordinate system, 106
    relative velocity, 108
    slip velocity, 107
    steady-state equations, 107
Mixture theory, 105
Molecular imaging, 123–124, 141–142
Multiphase flow
    definition, 102
    flow map, 102–103
    flow regimes, 102–103

N

Nanofluidics, 187–188
Nanofluids
    AlN/EG, 95
    boiling heat transfer and CHFs
        flow boiling heat transfer, 14–16
        nucleate pool boiling (*see* Nucleate
            pool boiling heat transfer)
        research needs, 22–23
    characteristics, 32
    classification, 3
    definition, 32
    features/characteristics, 1–2
    heat transfer technique
            (*see* Nanofluids heat transfer,
            fractal approach)
    nanoparticles, 2
    nanoparticles fractal analysis
        cumulative number, 35

fractal power law, 33–34
fractal theory and technique,
        35–36
probability density function, 36
SEM image, TiO_2 nanoparticles, 34
size distribution, Al_2O_3
        nanoparticles, 34
TEM image, Al_2O_3 nanoparticles, 34
preparation
    base fluids, 88
    dispersions, 89
    nanopowders, 88–89
    reproducibility and sample
            homogeneity, 90
    temperature cycling, 89
SiO_2/W, 95
specific heat, 9
surface tension, 8–9
thermal conductivity (*see* Thermal
        conductivity, nanofluids)
transport properties, 32
viscous properties
    CNT nanofluids, 7–8
    CuO–water nanofluids, 6–7
    water–Al_2O_3 nanofluids, 7
Nanofluids heat transfer, fractal
        approach
convective heat transfer (*see*
        Convective heat transfer,
        nanofluids)
effective thermal conductivity
    average diameter of nanoparticles,
            40–42
    equivalent thermal conductivity,
            39–40
    fractal dimension *versus*
            volumetric concentration, 41
    heat convection, Brownian
            nanoparticles, 36–37
    heat transfer coefficient, Stokes
            region, 38
    hydrodynamic boundary
            layer, 39
    negative and positive
            correlation, 43
    Prandtl number, 38
    quantity of heat, 38
    Reynolds number, 37
    silica nanoparticles in water, 43

suspended nanoparticles in base
  fluids, 36
total surface area, 39
pool boiling heat transfer (*see* Pool
  boiling heat transfer)
Nanoneedles, 194–195
Nanooncology
biomarker, 123–124
cancer management/detection
  (*see* Cancer management/
  detection)
contrast discrimination measures
  aqueous ʟ-phenylalanine, 155–158
  contrast metrics, 155
  glucose solution, 157, 159–161
  photothermal therapy, 161
definition, 122
fluid dynamics
  assumptions, 129
  bistable switch, 133
  drug transport, 132–133
  modeling framework, 129–130
  monostable switch, 134
  symbols definition, 129, 131
  tumor blood flow, 131–132
  tumor cell density dynamics, 134
  tumor microenvironment, 128–129
  tumor vessel, 128
fluid mechanics, 122
gold nanoparticles, 124
metabolomics, 123
molecular imaging, 123–124
nanoscale flow imaging (*see*
  Nanoscale flow imaging)
nanotechnology-based drug delivery
  systems, 134–135
  dendrimers, 139
  drug delivery, 137
  hydrophobic and hydrophilic
    drugs, 134
  liposomal formulation, 137–138
  nanoparticle drug complexes,
    135–136
  polymeric nanoparticles, 138–139
omics technologies, 124
Nanoscale flow imaging
chi-square distribution, 152–154
degree of freedom, 154
Gaussian-like distributions, 152–153

polarimetric fusion
  digital image distribution, 151
  DOLP, 150
  DOP, 151
  FOMs, 152
  MSE, 152
  SNR, 152
polarimetric principles, 149–150
Nanotechnology-based drug delivery
  systems, 134–135
dendrimers, 139
drug delivery, 137
hydrophobic and hydrophilic
  drugs, 134
liposomal formulation, 137–138
nanoparticle drug complexes, 135–136
polymeric nanoparticles, 138–139
Near-infrared (NIR) QDs, 149
Nucleate pool boiling heat transfer
active nucleation site density, 64
Al_2O_3–water nanofluids, 11–14
γ-Al_2O_3–water nanofluids, 12
bubble departure diameter, 65
bubble departure frequency, 65–66
bubble generation and departure, 64
CHFs (*see* Critical heat fluxes (CHFs))
CNTs, 12–13
CuO/water nanofluids, 11
vs. flow boiling heat transfer, 21–22
fractal model, 67
industrial applications, 2
liquid–solid contact angle, 64
natural convection, 66
refrigerant R-141b, 10–11
SiO_2–water nanofluids, 14
total heat flux, 64
ZrO_2–water nanofluids, 12

O

Ohm's law, 83
Omics technologies, 124

P

Periodic table, 234
Photon equations, 209
Photothermal therapy, 161
Physical constants, 208

Polymeric nanoparticles, 138–139
Pool boiling heat transfer
   nucleate pool boiling heat transfer
      active nucleation site density, 64
      bubble departure diameter, 65
      bubble departure frequency, 65–66
      bubble generation and
        departure, 64
      CHFs (*see* Critical heat fluxes
        (CHFs))
      fractal model, 67
      liquid–solid contact angle, 64
      natural convection, 66
      total heat flux, 64
   present model predictions *vs.*
      experimental data
      boiling characteristics, 68–69
      $CaCO_3$ nanofluids, 67
      Cu nanofluids, 67–68
      nanoparticle diameter with
        volume fractions, 69–71
      surface roughness, 68–69
      water-based nanosuspensions, 70
   subcooled pool boiling
      active cavity density and size, 62
      bubble departure frequency,
        61–62
      bubble waiting time, 61–62
      fractal analytical model, 60, 63
      heat flux equation, 60
      mechanisms, 59
      minimum diameter of active
        cavity, 63
      number of active cavities, 60
      volume of single bubble at
        departure, 60–61
Probability distributions table, 222–223

Q

Quantum dots (QDs)
   applications, 124
   breast cancer, 148
   cancer diagnosis, 124
   carbohydrate antigen 125, 147–148
   lymph nodes metastasis, 149
   optical and electrical properties, 140
   optical imaging, 143–144
   silk fibroin, 148

R

Refrigerant R-141b, 10–11

S

Signal-to-noise-ratio (SNR), 152
Silica (SiO_2) nanopowder, 88
Silicon nanowires (SiNWs), 146
Silk fibroin (SF), 148
Small quantities approximation,
   225–226
Solid lipid nanoparticles (SLNs), 137
Superparamagnetic iron oxide NPs
   (SPIONs), 144
Surface tension, 8–9

T

Thermal conductivity, nanofluids
   accuracy of, 83–84
   Brownian motion, 6
   CNT suspensions, 5
   effective medium theoretical models,
      95, 98
      composite layer, 96
      Hamilton, and Crosser model,
        96–97
      two-layer model, 96–97
   enhancement definition, 84
   experimental data, CuO nanofluids,
      93–95
   experimental setup, 84–87
      electrical resistance, platinum
        wires, 87–88
      heating curves, platinum wires,
        86–87
   heat transfer performance, 3–4, 6
   histograms, 90–91
   hot-wire calorimeter, 80–81
   linear variation of resistance, 83
   metallic nanofluids, 4
   nanoparticles fractal analysis, 33
   preparation of nanofluids, 88–90
   solution of heat equation, 81–82
   statistics, 90–91
   temperature increment, 82
   volume fraction of particles,
      92–93

Transient hot-wire technique, 81
Trigonometric identities, 216–217

V

Vectors
    derivatives, 231–232
    fundamental theorems, 233
    identities, 232–233

W

Water (W), nanofluid preparation,
    88–90
Wavelengths
    commercial lasers, 212
    light, optical ranges, 211
    photon equation, 209

Printed and bound by CPI Group (UK) Ltd, Croydon, CR0 4YY

22/10/2024

01777621-0011